Frederic P. Miller, Agnes F. Vandome,
John McBrewster (Ed.)

PlayStation 2

Frederic P. Miller, Agnes F. Vandome, John McBrewster (Ed.)

PlayStation 2

PlayStation 2. PSX (DVR), DualShock, PlayStation 2 Expansion Bay, EyeToy, PlayStation 2 Headset, Logitech, Linux for PlayStation 2, PCSX2, List of PlayStation 2 games

Alphascript Publishing

Imprint

Publisher:
Alphascript Publishing is a trademark of
VDM Publishing House Ltd.,17 Rue Meldrum, Beau Bassin,1713-01 Mauritius

Email: info@vdm-publishing-house.com
Website: www.vdm-publishing-house.com

Published in 2009

Printed in: U.S.A., U.K., Germany. This book was not produced in Mauritius.

ISBN: 978-613-0-07717-4

Contents

Articles

References

PlayStation 2

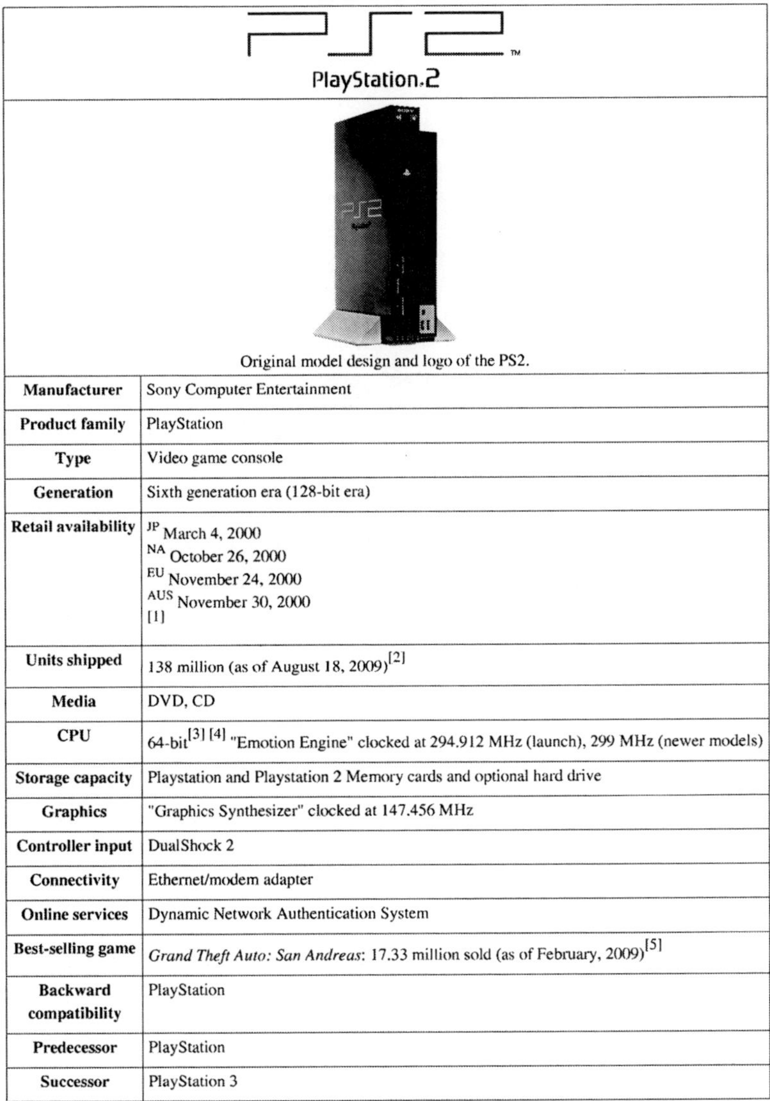

Original model design and logo of the PS2.

Manufacturer	Sony Computer Entertainment
Product family	PlayStation
Type	Video game console
Generation	Sixth generation era (128-bit era)
Retail availability	JP March 4, 2000 NA October 26, 2000 EU November 24, 2000 AUS November 30, 2000 [1]
Units shipped	138 million (as of August 18, 2009)[2]
Media	DVD, CD
CPU	64-bit[3] [4] "Emotion Engine" clocked at 294.912 MHz (launch), 299 MHz (newer models)
Storage capacity	Playstation and Playstation 2 Memory cards and optional hard drive
Graphics	"Graphics Synthesizer" clocked at 147.456 MHz
Controller input	DualShock 2
Connectivity	Ethernet/modem adapter
Online services	Dynamic Network Authentication System
Best-selling game	*Grand Theft Auto: San Andreas*: 17.33 million sold (as of February, 2009)[5]
Backward compatibility	PlayStation
Predecessor	PlayStation
Successor	PlayStation 3

The **PlayStation 2** (often shortened to **PS2**) is a sixth-generation video game console manufactured by Sony. The successor to the PlayStation, and the predecessor to the PlayStation 3, the PlayStation 2 forms part of the PlayStation series of video game consoles. Its development was announced in March 1999 and it was released a year later in Japan. Its primary competitors were Sega's Dreamcast, Microsoft's Xbox, and Nintendo's GameCube.

The PS2 is the best-selling console to date, having reached over 138 million units sold as of August 18, 2009[2] and a software library projected to exceed 1,900 games in 2009.[6] Twenty games are scheduled to be released in 2010, giving the PS2 a marketable life of over 10 years, thus continuing the sixth generation.

History

Only a few million people had obtained consoles by the end of 2000 due to manufacturing delays.[7] Directly after its release, it was difficult to find PS2 units on retailer shelves.[8] Another option was purchasing the console online through auction websites such as eBay, where people paid over one thousand dollars for a PS2.[9] The PS2 initially sold well partly on the basis of the strength of the PlayStation brand and the console's backward compatibility, selling over 980,000 units in Japan by March 5, 2000, one day after launch.[10] This allowed the PS2 to tap the large install base established by the PlayStation — another major selling point over the competition. Later, Sony added new development kits for game developers and more PS2 units for consumers.

A notable piece of advertising for the PS2 launch was accompanied by the popular "PS9" television commercial. 9 was to be the epitome of development, toward which the PS2 was the next step. The ad also presaged the development of the PlayStation PortableWikipedia:Citation needed (first released in Japan on December 12, 2004).

Many analysts predicted a close three-way matchup between the PS2 and competitors Microsoft's Xbox and the Nintendo GameCube (GameCube being the cheapest of the three consoles and had an open market of games); however, the release of several blockbuster games during the 2001 holiday season maintained sales momentum and held off the PS2's rivals.[11]

Although Sony, unlike Sega with its Dreamcast, placed little emphasis on online gaming during its first yearsWikipedia:Citation needed, that changed upon the launch of the online-capable Xbox. Sony released the PlayStation Network Adaptor in late 2002 to compete with Microsoft, with several online first–party titles released alongside it, such as *SOCOM: U.S. Navy SEALs* to demonstrate its active support for Internet play. Sony also advertised heavily, and its online model had the support of Electronic Arts. Although Sony and Nintendo both started out late, and although both followed a decentralized model of online gaming where the responsibility is up to the developer to provide the servers, Sony's attempt made online gaming a major selling point of the PS2.

In September 2004, in time for the launch of *Grand Theft Auto: San Andreas*, Sony revealed a new, slimmer PS2 (see *Hardware revisions*). In preparation for the launch of the new model (SCPH-70000), Sony stopped making the older model (SCPH-5000x) to let the distribution channel empty its stock of the units.Wikipedia:Citation needed After an apparent manufacturing issue—Sony reportedly underestimated demand—caused some initial slowdown in producing the new unit caused in part by shortages between the time the old units were cleared out and the new units were ready. The issue was compounded in Britain when a Russian oil tanker became stuck in the Suez Canal, blocking a ship from China carrying PS2s bound for the UK. During one week in November, British sales totalled 6,000 units — compared to 70,000 units a few weeks prior.[12] There were shortages in more than 1700 stores in North America on the day before Christmas.[13]

On November 29, 2005, the PS2 became the fastest game console to reach 100 million units shipped, accomplishing the feat within 5 years and 9 months from its launch. This achievement occurred faster than its predecessor, the PlayStation, which took 9 years and 6 months to reach the same benchmark.[14]

Sony has announced that starting April 1, 2009 the PS2 would be retailing at the new price of $99.99.[15]

Hardware and software compatibility

See also: List of PlayStation 2 games, List of PlayStation games incompatible with PlayStation 2, List of PlayStation 2 CD-ROM games, List of PlayStation 2 DVD-9 games, List of PlayStation 2 games with HD support, and Chronology of PlayStation 2 games

OEM PlayStation 2 8 MB Memory Card.

In addition to PS2 software, the PS2 can read both CDs and DVDs and is backward compatible with PlayStation games. The PS2 also supports PlayStation memory cards (for PlayStation game saves only) and controllers, although the memory cards only work with PS1 games and the controllers may not support all functions (such as analog buttons) for PS2 games.

The PlayStation 2's DualShock 2 controller is cosmetically similar to the original DualShock.

The PS2's → DualShock 2 controller is essentially an upgraded PlayStation DualShock; analog face, shoulder and D-pad buttons replaced the digital buttons of the original. Like its predecessor, the DualShock 2 controller has force feedback, which is commonly called the "vibration" function. The standard PlayStation 2 memory card has an 8MB capacity and uses Sony's MagicGate encryption. This requirement prevented the production of memory cards by third parties who did not purchase a license for the MagicGate encryption. Memory cards without encryption can be used to store PlayStation game saves, but PlayStation games would be unable to read from or write to the card - such a card could only be used as a backup.

The console also features USB and IEEE 1394 expansion ports. Compatibility with USB and IEEE 1394 devices is dependent on the software supporting the device. For example, the PS2 BIOS will not boot an ISO image from a USB flash drive or operate a USB printer, as the machine's operating system does not include this functionality. By contrast, *Gran Turismo 4* is programmed to save screenshots to a USB mass storage device and print images on certain USB printers. A PlayStation 2 HDD can be installed in an → expansion bay on the back of the console, with some exceptions (see *Hardware revisions* below).

Online

See also: List of PlayStation 2 network games

PlayStation 2 infrared remote control

With the purchase of a separate unit called the Network Adapter (which is built into the slimline model), some PS2 games support online multiplayer. Instead of having a unified, subscription-based online service like Xbox Live, online multiplayer on the PS2 is split between publishers and run on third-party servers. Most recent PS2 online games have been developed to exclusively support broadband Internet access. Xbox Live similarly requires a broadband Internet connection.

All online PS2 games released in and after 2003 are protected by the Dynamic Network Authentication System (DNAS). The purpose of this system is to prevent piracy and online cheating. DNAS will prevent games from being played online if they are determined to be pirated copies or if they have been modified. However, methods have been developed to get around this protection by modifying key files in the modified game.

Also, some unofficial modifications have been made on the PS2 software allowing it to be used as a fully-functional web browser or messenger when connecting to a certain network. The PS2 can also run Linux.

Hardware revisions

The PS2 has undergone many revisions, some only of internal construction and others involving substantial external changes. These are colloquially known among PS2 hardware hackers as V0, V1, V2, *etc.*, up to V15b[16] (as of 2008).

The PS2 is primarily differentiated between models featuring the original case design and "slimline" models, which were introduced at the end of 2004.

Original case design

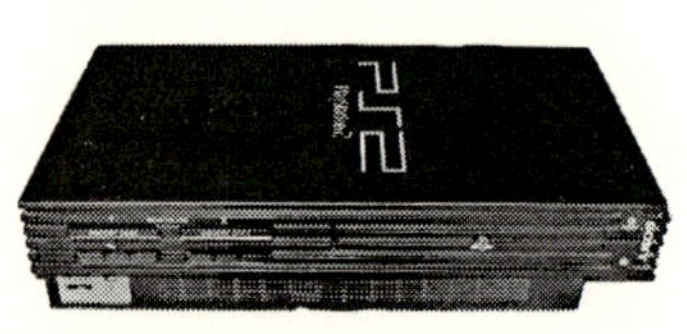
The original PlayStation 2 design.

Three of the original PS2 launch models (SCPH-10000, SCPH-15000, and SCPH-18000) were only sold in Japan, and lacked the expansion bay (Dev9) of current PS2 models. These models included a PCMCIA slot instead of the Dev9 port of newer models. A PCMCIA-to-Dev9 adapter was later made available for these modelsWikipedia:Citation needed. SCPH-10000 and SCPH-15000 did not have a built-in DVD movie playback and instead relied on encrypted playback software that was copied to a memory card from an included CD-ROM. (Normally, the PS2 will only execute encrypted software from its memory card, but see PS2 Independence Exploit.) V3 had a substantially different internal structure from the subsequent revisions, featuring several interconnected printed circuit boards. As of V4 everything was unified into one board, except the power supply. V5 introduced minor internal changes, and the only difference between V6 (sometimes called V5.1) and V5 is the orientation of the Power/Reset switch board connector, which was reversed to prevent the use of no-solder modchips. V7 and V8 included only minor revisions to V6. Assembly of the PS2 moved to the People's Republic of China during the development of V9 (model numbers SCPH-50000 and SCPH-50001). The upgraded console added an infrared port for the optional DVD remote control, removed the IEEE 1394 port, added the capability to read DVD-RW and DVD+RW discs, added progressive-scan output of DVD movies, and added a quieter fan. V10 and V11 were only minor revisions to V9.

The PS2 standard color is matte black. Several different variations in color have been produced in different quantities and regions, including ceramic white, light yellow, metallic blue (aqua), metallic silver, navy (star blue), opaque blue (astral blue), opaque black (midnight black), pearl white, Sakura purple, satin gold, satin silver, snow white, super red, and transparent blue (ocean blue).[17] [18] [19] [20]

The small PlayStation logo on the front of the disc tray could be rotated ninety degrees, in order for the logo to be the right way up in both vertical and horizontal console orientations.

Slimline

In September 2004, Sony unveiled its third major hardware revision (V12, model number SCPH-70000). Available in late October 2004, it is smaller, thinner, and quieter than the older versions and includes a built-in Ethernet port (in some markets it also has an integrated modem). Due to its thinner profile, it does not contain the 3.5" → expansion bay and therefore does not support the internal hard disk drive. It also lacks an internal power supply, similar to the GameCube, and has a modified Multitap expansion. The removal of the expansion bay has been criticized as a limitation due to the existence of titles such as *Final Fantasy XI*, which require the use of the HDD. The official → PS2 Linux also requires an expansion bay to function. Currently only the modified Multitap is sold in stores, meaning that owners of older PS2s must find a used or non-Sony Multitap in order to have 4 or 8 players during a single game. Third-party connectors can be soldered into the unit giving hard drive support, however IDE connections were completely removed in the V14 revision, thereby eliminating this option. Certain mod chips enable the use of a USB hard drive or other mass storage device.

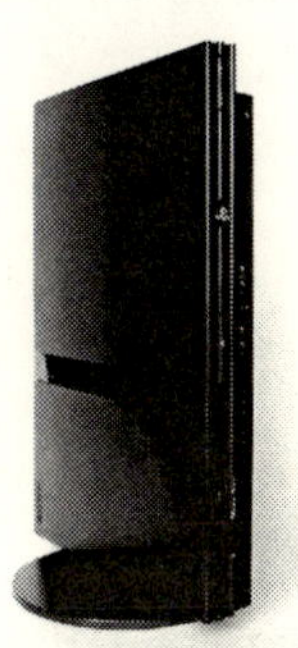

The redesigned slimline PlayStation 2.

The redesigned slimline PlayStation 2 in silver.

Comparison of the slimline PlayStation 2 design with the PlayStation 2, with an Eye Toy on top.

There are some disputes on the numbering for this PS2 version, Wikipedia:Citation needed since there are actually two sub-versions of the SCPH-70000.[21] One of them includes the old EE and GS chips, and the other contains the newer unified EE+GS chip, but otherwise they are identical. Since the V12 version had already been established for this model, there were some disputes regarding these sub-versions. Two propositions were to name the old model (with separate EE and GS chips) V11.5 and the newer model V12, and to name the old model V12 and the newer model V13. Currently, most people use V12 for both models, or V12 for the old model and V13 for the newer one.

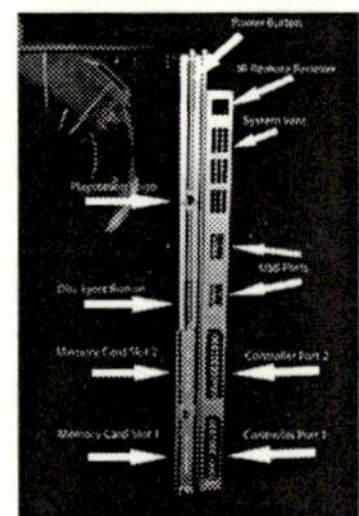

Slimline PS2 with its connectors labeled in the front.

The V12 model was first released in black, but a silver edition is available in the United Kingdom, Germany, Australia, United Arab Emirates and other GCC Countries, France, Italy, South Africa, and most recently, North America. It is unknown whether or not this will follow the color schemes of the older model, although a limited edition console that is pink in colour has become available since March 2007.

V12 (or V13) was succeeded by V14 (SCPH-75001 and SCPH-75002), which contains integrated EE and GS chips, and different ASICs compared to previous revisions, with some chips having a copyright date of 2005, compared to 2000 or 2001 for earlier models. It also has a different lens and some compatibility issues with a different number of PlayStation games and even some PS2 games.

Silver slimline PS2 with its connectors labeled in the back

In the beginning of 2005 it was found that some black slimline console power transformers bought between November and December 2004 were faulty and could overheat. The units were recalled by Sony, with the company supplying a replacement model made in 2005.

Later hardware revisions had better compatibility with PlayStation games (*Metal Gear Solid: VR Missions* operates on most silver models); however, the new Japanese slim models have more issues with playing PlayStation games than the first PS2 revisions. Wikipedia:Citation needed

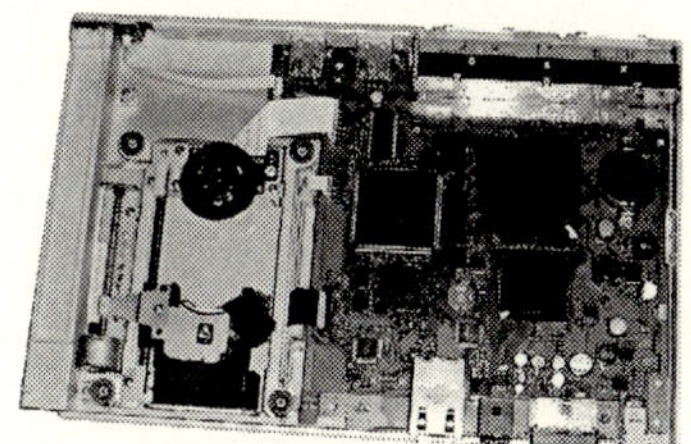
The mainboard of the silver slimline PS2 (model SCPH-79000).

In 2006, Sony released new hardware revisions (V15, model numbers SCPH-77001a and SCPH-77001b). It was first released in Japan on September 15, 2006, including the Silver edition. After its release in Japan, it was then released in North America, Europe, and other parts of the world. The new revision uses an integrated, unified EE+GS chip, a redesigned ASIC, a different laser lens, an updated BIOS, and updated drivers.

In July 2007, Sony started shipping a revision of the slimline PlayStation 2 (SCPH-79000) featuring a reduced weight of 600 grams compared to 900 grams of the SCPH-77001, achieved through a reduction in parts. The unit also uses a smaller motherboard as well as a custom ASIC which houses the Emotion Engine, Graphics Synthesizer, and the RDRAM. The AC adaptor's weight was also reduced to 250 grams from the 350 grams in the previous revision.[22]

Another refinement of the slimline PlayStation 2 (SCPH-90000) was released in Japan on November 22, 2007[23] , and in the US and EU in late 2008Wikipedia:Citation needed, with an overhauled internal design that incorporates the power supply into the console itself, with a further reduced total weight of 720 grams. SCPH-90000 series consoles manufactured after March 2008 incorporate a revised BIOS, which disables an exploit present in all older models that allowed homebrew applications to be launched from a memory card.

PSX

Main article: → PSX (DVR)

Sony also manufactured a consumer device called the → PSX that can be used as a digital video recorder and DVD burner in addition to playing PS2 games. The device was released in Japan on December 13, 2003 and though it was never released any where else, it can be found for sale in some of the Sony Style shops located in a number of countries. The PSX was poorly received in both areas, some major features were absent from the first revisions of the hardware and experienced very weak sales in spite of major price drops. The system is considered a rarity and is now selling for around $500 on eBay. The PSX (DVR) was also the first Sony product to include the XrossMediaBar interface.[24]

Sales

Region	Units sold	First available
Japan	21 million (as of October 1, 2008)[25]	March 4, 2000
North America	50 million (as of December 2008)[26]	October 26, 2000
Europe	48 million (as of May 6, 2008)[27]	November 24, 2000
Worldwide	138 million (as of August 18, 2009)[2]	

On November 29, 2005, the PlayStation 2 became the fastest game console to reach 100 million units shipped, accomplishing the feat within 5 years and 9 months from its launch. This achievement occurred faster than its predecessor, the PlayStation, which took 9 years and 6 months to reach the same benchmark.[14]

The PS2 has sold 138 million sell-in units worldwide as of August 18, 2009, according to Sony.[2] In Europe, the PS2 has sold 48 million units as of May 6, 2008, according to Sony Computer Entertainment Europe.[27] In North America, the PS2 has sold 50 million units as of December 2008.[26] In Japan, the PS2 has sold 21,454,325 units as of October 1, 2008, according to *Famitsu*/Enterbrain.[25]

In Europe, the PS2 sold 6 million units in 2006 and 3.8 million in 2007, according to estimates by Electronic Arts.[28] [29] In 2007, the PS2 sold 3.97 million units in the US according to the NPD Group[30] [31] and 816,419 units in Japan according to Enterbrain.[32] In 2008, the PS2 sold 480,664 units in Japan, according to Enterbrain.[32] [33]

Accessories

Main articles: → DualShock, PlayStation 2 HDD, → EyeToy, → PlayStation 2 Headset, and → Logitech

The PlayStation 2's DualShock 2 controller is largely identical to the PlayStation's DualShock, with the same basic functionality; however, it includes analog pressure sensitivity on the face, shoulder and D-pad buttons, is lighter and includes two more levels of vibration. The L2 and R2 buttons are also significantly larger. The fact that the design did not change pleased some consumers who were already used to the DualShock controller.Wikipedia:Citation needed

The EyeToy digital camera sitting atop a Slimline PS2.

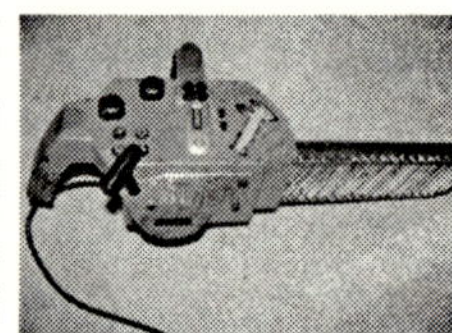

Resident Evil 4 chainsaw controller compatible with the PlayStation 2.

Optional hardware includes DualShock or DualShock 2 controllers, a PS2 DVD remote control, an internal or external HDD, a network adapter, horizontal and vertical stands, PlayStation or PS2 memory cards, light guns (GunCon), fishing rod and reel controllers. Also available are various cables and interconnects, including the Multitap for PlayStation or PS2, S-Video, RGB, SCART, VGA (for progressive scan games and PS2 Linux only), component and composite video cables, an RF modulator, a USB camera (→ EyeToy), dance pads for *Dance Dance Revolution*, *In the Groove*, and *Pump It Up* titles, Konami microphones for use with the *Karaoke Revolution* games, dual microphones (sold with and used exclusively for *SingStar* games), various "guitar" controllers (for the *Guitar Freaks* series and *Guitar Hero* series), the drum set controller (sold in a box set (or by itself) with a "guitar" controller and a USB microphone for use with *Rock Band*), *Onimusha 3* katana controller, *Resident Evil 4* chainsaw controller, a USB keyboard and mouse, and a headset. Unlike the PlayStation, which required the use of an official Sony PlayStation mouse to play mouse-compatible games, the few PS2 games with mouse support work with standard PC-compatible USB mice. Early versions of the PS2 could be networked via an iLink port, though this had little game support and was dropped. The original PS2 multitap cannot be plugged into the newer slim models (as the multitap connects to the memory card slot as well as the controller slot and the memory card slot on the slimline is shallower). New slim-design multitaps are manufactured for these models, however third-party adapters also exist to permit original multitaps to be used. Some third party manufacturers have created devices that allow disabled people to access the PS2 through ordinary switches etc. One such device is the PS2-SAP from LEPMIS, another is for example the JPemulator.

Homebrew development

See also: Linux for PlayStation 2

Sony released a version of the Linux operating system for the PS2 in a package that also includes a keyboard, mouse, Ethernet adapter and HDD. Currently, Sony's online store states that the Linux kit is no longer for sale in North America. However as of July 2005, the European version was still available. The kit boots by installing a proprietary interface, the run-time environment, which is on a region-coded DVD, so the European and North America kits only work with a PS2 from their respective regions.

In Europe and Australia, the PS2 comes with a free Yabasic interpreter on the bundled demo disc. This allows simple programs to be created for the PS2 by the end-user. This was included in a failed attempt to circumvent a UK tax by defining the console as a "computer" if it contained certain software.Wikipedia:Citation needed

A port of the NetBSD project and BlackRhino GNU/Linux, an alternative Debian-based distribution, are also available for the PS2.

Using homebrew programs (e.g. 'SMS Media Player'[34]) it is possible to listen to various audio file formats (MP3, OMA, Ogg Vorbis, AAC, FLAC, AC3), and watch various video formats (DivX/XviD, MPEG1, MPEG2, MPEG4-ASP in AVI Container) using the console. Media can be played from any device connected to the console i.e. external USB/Firewire thumb drive/hard disk (FAT32 only), the internal hard disk on early revision consoles, optical CD-R(W)/DVD±R(W) disks (modded systems or patched disks), or network shares (Windows Network or PS2 host: protocol).

Homebrew programs can be launched directly from a memory card on unmodified consoles by using certain software that takes advantage of a long known and used exploit, dealing with the boot part of the EE/IOP process (Independence). A recent development (May 2008) allows homebrew programs to be launched without a trigger disc such as is needed in the older exploit, which also allows use of homebrew on unmodded systems with a dead disc drive (Free McBoot), however installation of the exploit to each individual memory card (copying does NOT work) requires an already exploited/modded system in order to launch the installer, this newer exploit will not work on the very newest PS2s (SCPH-9000x model with BIOS 2.30 and up) but will work on ALL models prior to that, including slimlines.

Homebrew programs can be used to play patched backups of original PS2 DVD games on unmodified consoles, and to install retail discs to an installed hard drive on older models (ESR, HDLoader, USBAdvance). This is illegal in many countries.

Homebrew emulators of older computer and gaming systems have been developed for the PS2[35] . Using these homebrew programs the PS2 can emulate the following: Atari 2600, Atari 5200, BBC Micro, Commodore 64, Game Boy, Sega Mega Drive, Sega Master System, MSX, Neo Geo, Nintendo Entertainment System, TurboGrafx-16, and Super Nintendo Entertainment System.

Technical specifications

The specifications of the PlayStation 2 console are as follows, with hardware revisions:

- CPU: 64-bit[3] [4] "Emotion Engine" clocked at 294.912 MHz (299 MHz on newer versions), 10.5 million transistors
  - System Memory: 32 MB(32 × 2^{20} bytes) Direct Rambus or RDRAM
  - Memory bus Bandwidth: 3.2 gigabytes per second
  - Main processor: MIPS R5900 CPU core, 64 bit, little endian (mipsel).
  - Coprocessor: FPU (Floating Point Multiply Accumulator × 1, Floating Point Divider × 1)
  - Vector Units: VU0 and VU1 (Floating Point Multiply Accumulator × 9, Floating Point Divider × 1), 32-bit, at 150 MHz.
    - VU0 typically used for polygon transformations optionally (under parallel or serial connection), physics and other gameplay based things
      - Parallel performs transformations in parallel in the same moment
      - Serial (series) performs transformations in a series of steps or stages coherent to the design of each VU
        - Stage 1: VU0 does perspective and cam, boning, animations and movement laws per triangle
        - Stage 2: VU1 does colors, lights and effects per triangle)
    - VU1 typically used for polygon transformations, lighting and other visual based calculations
      - Texture matrix able for 2 units (UV/ST)[36]
  - Floating Point Performance: 6.2 gigaFLOPS (single precision 32-bit floating point)
    - FPU 0.64 gigaFLOPS
    - VU0 2.44 gigaFLOPS
    - VU1 3.08 gigaFLOPS (with Internal 0.64 gigaFLOP EFU)
  - 3D CG Geometric transformation(VU0+VU1 parallel): 66 million polygons per second
    - 3D CG Geometric transformations under curved surfaces: 16 million polygons per second
    - 3D CG Geometric transformations at peak bones/movements/effects(textures)/lights(VU0+VU1): 15-20 million polygons per second (dependent on if series or parallel T&L)
    - Actual real-world polygons (per frame):500-650k at 30fps, 250-325k at 60fps
  - Compressed Image Decoder: MPEG-2

Emotion Engine CPU

Graphics Synthesizer GPU

Graphics Synthesizer as on SCPH39000.

Older EE+GS that does not incorporate system memory (Found in Older Charcoal Black Slim PS2s. (SCPH-70001).

- I/O Processor interconnection: Remote Procedure Call over a serial link, DMA controller for bulk transfer
- Cache memory: Instruction: 16 KB(16 x 2^{10} bytes), Data: 8 KB + 16 KB (ScrP)

- Graphics processing unit: "Graphics Synthesizer" clocked at 147 MHz
  - Pixel pipelines: 16
  - Video output resolution: variable from 256x224 to 1280x1024 pixels
  - 4 MB (4 x 2^{20} bytes) Embedded DRAM video memory bandwidth at 48 gigabytes per second (main system 32 MB can be dedicated into VRAM for off-screen materials)
    - Texture buffer bandwidth: 9.6 GB/s
    - Frame buffer bandwidth: 38.4 GB/s
  - DRAM Bus width: 2560-bit (composed of three independent buses: 1024-bit write, 1024-bit read, 512-bit read/write)
  - Pixel Configuration: RGB: Alpha:Z Buffer (24:8, 15:1 for RGB, 16, 24, or 32-bit Z buffer)
  - Dedicated connection to: Main CPU and VU1
  - Overall Pixel fillrate: 16x147 = 2.352 Gpixel/s (rounded to 2.4 Gpixel/s)
    - Pixel fillrate: with no texture, flat shaded 2.4(75,000,000 32pixel raster triangles)
    - Pixel fillrate: with 1 full texture(Diffuse Map), Gouraud shaded 1.2 (37,750,000 32-bit pixel raster triangles)
    - Pixel fillrate: with 2 full textures(Diffuse map + specular or alpha or other), Gouraud shaded 0.6 (18,750,000 32-bit pixel raster triangles)
  - GS effects: AAx2 (poly sorting required)[37], Bilinear, Trilinear, Multi-pass, Palletizing (4-bit = 6:1 ratio, 8-bit = 4:1)
  - Multi-pass rendering ability
    - Four passes = 300 Mpixel/s (300 Mpixels/s divided by 32 pixels = 9,375,000 triangles/s lost every four passes)[38]

ASIC that incorporates the EE, GS, and system memory (found in silver slim PS2s. Model SCPH-79000).

- Audio: "SPU1+SPU2" (SPU1 is actually the CPU clocked at 8 MHz)
  - Number of voices: 48 hardware channels of ADPCM on SPU2 plus software-mixed channels
  - Sampling Frequency: 44.1 kHz or 48 kHz (selectable)
  - Output: Dolby Digital 5.1 Surround sound, DTS (Full motion video only), later games achieved analog 5.1 surround during gameplay through Dolby Pro Logic II

The I/O Processor, containing a MIPS R3000-based CPU used in the PlayStation 1 and I/O logic

- I/O Processor
  - CPU Core: Original PlayStation CPU (MIPS R3000A clocked at 33.8688 MHz or 37.5 MHz)
  - Automatically underclocked to 33.8688 MHz to achieve hardware backwards compatibility with original Playstation format games.

- Sub Bus: 32-bit
- Connection to: SPU and CD/DVD controller.

- Interfaces:
  - 2 proprietary PlayStation controller ports (250 kHz clock for PS1 and 500 kHz for PS2 controllers)
  - 2 proprietary Memory Card slots using MagicGate encryption (250 kHz for PS1 cards, up to 2 MHz for PS2 cards)
  - Expansion Bay (PCMCIA on early models for PCMCIA Network Adaptor and External Hard Disk Drive) DEV9 port for Network Adaptor
  - Modem, Ethernet and Internal Hard Disk Drive (single IDE/ATA channel, possible to hook 2 devices to.)
  - FireWire (only in SCPH 10xxx – 3xxxx)
  - Infrared remote control port (SCPH 5000x and newer) — *IEEE 1394 port removed and Infrared port added in SCPH-50000 and later hardware versions.*
  - 2 USB 1.1 ports with an OHCI-compatible controller.
- Disc Drive type: proprietary interface through a custom micro-controller + DSP chip. 24x speed (PlayStation 2 format CD-ROM, PlayStation format CD-ROM), 4x (Supported DVD formats) — Region-locked with anti-copy protection. Can't read "Gold Discs" *i.e.*, normal CD-ROMs.
- Supported Disc Media: PlayStation 2 format CD-ROM, PlayStation format CD-ROM, Compact Disc Audio, PlayStation 2 format DVD-ROM (4.7 GB)(some games on DVD9 8.5 GB), DVD Video (4.7 GB), DVD-9 (8.5 GB Double-Layer). Later models (starting with SCPH-50000) are DVD+RW, and DVD-RW compatible.

Emulation

See also: → PCSX2

See also

- HDD Utility Disc
- PlayStation Broadband Navigator

External links

Official sites

- PlayStation official site [39]
- Official PS2 Developer Site – only for registered developers [40]

Directories

- PlayStation 2 [41] at the Open Directory Project

References

[1] Sony: 120 million PS2s sold - News at GameSpot (http://www.gamespot.com/news/6181828.html)

[2] Plunkett, Luke (2009-08-18). " Sony Talk PlayStation Lifetime Sales, PSN Revenue (http://kotaku.com/5340392/sony-talk-playstation-lifetime-sales-psn-revenue)". Kotaku. . Retrieved 2009-08-19.

[3] John L. Hennessy and David A. Patterson. "Computer Architecture: A Quantitative Approach, Third Edition". ISBN 1-55860-724-2

[4] Keith Diefendorff. "Sony's Emotionally Charged Chip". Microprocessor Report, Volume 13, Number 5, April 19, 1999. Microdesign Resources.

[5] Guinness World Records, ed. *Guinness World Records 2009 Gamer's Edition*. p. 108-109. ISBN 1904994459. "*GTA: San Andreas* is the best-selling PlayStation 2 game of all time, with a massive 17.33 million copies sold."

[6] Graft, Kris (March 31, 2008). " Sony: 250 PS2 "Greatest Hits" (http://www.edge-online.com/news/sony-250-ps2-greatest-hits)". Edge. . Retrieved April 29, 2009.

[7] " PS2 history (http://www.gamesindustry.biz/content_page.php?aid=21241)". gamesindustry.biz. 2006-11-22. . Retrieved 2007-09-25.

[8] " PlayStation 2 Timeline (http://archive.gamespy.com/articles/february04/ps2timeline/index2.shtml)". GameSpy. pp. 2-3. . Retrieved 2007-12-19.

[9] " PlayStation 2 Timeline (http://archive.gamespy.com/articles/february04/ps2timeline/index3.shtml)". GameSpy. pp. 3. . Retrieved 2007-12-19.

[10] " PlayStation 2 Timeline (http://archive.gamespy.com/articles/february04/ps2timeline/index2.shtml)". GameSpy. pp. 2. . Retrieved 2007-12-19.

[11] Chris Morris (May 14, 2002). " Sony slashes PlayStation prices: Preemptive move undercuts competition and could spark video game price war (http://money.cnn.com/2002/05/14/technology/ps2_pricecuts/)". CNN. .

[12] Valerie Elliott (December 9, 2004). " Merry Christmas, your PlayStation 2 is stuck in Suez (http://www.timesonline.co.uk/tol/news/uk/article400887.ece)". *Times Online* (News International). .

[13] " 2004 Holiday Sales Results Call (http://www.corporate-ir.net/ireye/ir_site.zhtml?ticker=GME&script=1010&item_id=997399)". GameStop. . Retrieved 2006-09-09.

[14] Sony Computer Entertainment (2005-11-30). " PlayStation 2 Breaks Record as the Fastest Computer Entertainment Platform to Reach Cumulative Shipment of 100 Million Units (http://www.scei.co.jp/corporate/release/pdf/051130e.pdf)". Press release. . Retrieved 2008-06-08.

[15] Keston, Lou (March 31, 2009). " Sony cuts price of older PlayStations to $100 (http://www.businessweek.com/ap/financialnews/D978VLNG0.htm)". Business Week. . Retrieved March 31, 2009.

[16] " PS2 Revision Identification (http://www.mod-chip.com/en/ps2version.htm)". Mod-Chip.com. .

[17] Calvert, Justin (2003-11-04). " PS2 price drop, new colors for Japan (http://uk.gamespot.com/ps2/action/mobilesuitzgundamaeugvstitans/news.html?sid=6078108)". GameSpot. . Retrieved 2007-07-10.

[18] Fahey, Rob (2004-03-09). " Sony launches new PS2 colours in Japan (http://www.gamesindustry.biz/content_page.php?aid=3080)". gamesindustry.biz. . Retrieved 2007-07-10.

[19] A list of all console colors and console Limited Editions (http://www.consolecolors.com/sony.html). Consolecolors.com

[20] Martyn Williams (October 14, 2001). " Sony: New hues on PlayStation 2 (http://archives.cnn.com/2001/TECH/fun.games/10/14/ps2.colors.idg/index.html)". CNN. . Retrieved 2007-07-10.

[21] Mod-Chip.Com - Your Primary Source Of Modchips For Your Console (http://www.mod-chip.com/ps2version.htm)

[22] Gantayat, Anoop (2007-06-08). " PS2 Gets Lighter (http://ps2.ign.com/articles/802/802262p1.html)". IGN.com. . Retrieved 2007-07-10.

[23] PLAYSTATION2 (SCPH-90000 SERIES) comes in a new design and in three color variations | Press releases | Sony Computer Entertainment Inc (http://www.scei.co.jp/corporate/release/071106ae.html)

[24] Fahey, Rob (2004-09-06). " Japanese retailers slash PSX prices as sales remain slow (http://www.gamesindustry.biz/content_page.php?aid=4280)". gamesindustry.biz. . Retrieved 2007-07-10.

[25] James Brightman (2008-10-20). " Xbox 360 Growth in Japan Has Topped All Platforms from March to September (http://www.gamedaily.com/articles/news/xbox-360-growth-in-japan-has-topped-all-platforms-from-march-to-september)". *GameDaily*. AOL. . Retrieved 2008-10-25.

[26] John Koller (January 15, 2009). " PS2 Sells Over 50 Million Units in North America, Breaks Console Sales Record (http://blog.us.playstation.com/2009/01/15/ps2-sells-over-50-million-units-in-north-america-breaks-console-sales-record/)". PlayStation.Blog. . Retrieved January 17, 2009.

[27] Ellie Gibson (2008-05-06). " PS3 has outsold Xbox 360 in Europe (http://www.eurogamer.net/article.php?article_id=137142)". Eurogamer. . Retrieved 2008-10-25.

[28] Electronic Arts (2008-01-31). " Supplemental Segment Information (http://media.corporate-ir.net/media_files/IROL/88/88189/Q3FY08SupSeg.pdf#page=4)". *Thomson Financial*. pp. 4. . Retrieved 2008-02-09.

[29] David Jenkins (2008-02-01). " EA Reveals European Hardware Estimates (http://www.gamasutra.com/php-bin/news_index.php?story=17206)". Gamasutra. . Retrieved 2008-02-09.

[30] James Brightman (2008-01-17). " NPD: U.S. Video Game Industry Totals $17.94 Billion, Halo 3 Tops All (http://www.gamedaily.com/articles/news/npd-us-video-game-industry-totals-1794-billion-halo-3-tops-all/19119/?biz=1)". GameDaily. . Retrieved 2008-08-02.

[31] Brandon Boyer (2008-01-18). " NPD: 2007 U.S. Game Industry Growth Up 43% To $17.9 Billion (http://www.gamasutra.com/php-bin/news_index.php?story=17006)". Gamasutra. . Retrieved 2008-08-02.

[32] " Japanese 2008 Market Report (http://www.mcvuk.com/interviews/403/JAPANESE-2008-MARKET-REPORT)". Market for Home Computing and Video Games. 2009-01-09. . Retrieved 2009-01-15.

[33] " 2008年国内ゲーム市場規模は約5826億1000万円（エンターブレイン調べ） (http://www.famitsu.com/game/news/1221045_1124.html)" (in Japanese). *Famitsu*. Enterbrain. 2009-01-05. . Retrieved 2009-01-15.

[34] Simple Media System for Playstation 2 - http://home.casema.nl/eugene_plotnikov/

[35] PS2 Emulators - http://www.sksapps.com/index.php?page=emus.html

[36] Power of PS2 (See data handling) (http://www.technology.scee.net/files/presentations/agdc2000/ThePowerOfPS2.pdf)

[37] Power of PS2 (See GS effects) (http://www.technology.scee.net/files/presentations/agdc2000/ThePowerOfPS2.pdf)

[38] PS2 Programming Optimisations (http://www.technology.scee.net/files/presentations/agdc2002/PS2Optimisations.pdf)

[39] http://www.playstation.com/

[40] https://www.ps2-pro.com/

[41] http://www.dmoz.org/Games/Video_Games/Console_Platforms/Sony/PlayStation_2/

PSX (DVR)

Manufacturer	Sony Corporation
Type	Digital video recorder/video game console
Retail availability	JP December 13, 2003 NA Not set
Media	DVD-ROM DVD-R DVD-RW DVD-RAM
Storage capacity	Hard disk, memory card

PSX was a Sony digital video recorder with fully integrated PlayStation and → PlayStation 2 video game consoles. Since it was designed to be a general-function audiovisual device, it was marketed by the main Sony Corporation and was released only in Japan on December 13, 2003. It was the first Sony product to utilize the XrossMediaBar.

Features

The PSX is a fully functional digital video recorder with RF, S-Video and composite inputs. It is able to tune analog VHF and CATV. It comes with a remote control and can be also linked with a PlayStation Portable to transfer videos and music via USB ports. [1] It also features software for video, photo and audio editing. [2]

The PSX supports PlayStation and → PlayStation 2 gaming using PlayStation 2-based hardware, with Emotion Engine, Graphics Synthesizer and the I/O processor. It supports online game compatibility using an internal broadband adapter. Games that utilize the PS2 HDD (for example Final Fantasy XI) are supported as well. [3]

The PSX doesn't include a controller. A special ceramic-white, USB DualShock 2 was released separately [4] . However, original DualShocks are fully supported by two joystick ports in the back side and memory cards are also supported via port in the front side. [5]

The PSX is also known for its introduction of Sony's XMB graphical user interface, used later on PlayStation Portable, PlayStation 3, 2008 BRAVIA model TVs and other Sony devices.

Models

The PSX XMB

The DESR-5000 and DESR-7000 were launch models and featured a 160 GB and 250 GB HDD. Later improved DESR-7000 models were launched in 2004 with bigger hard disk and supporting improved vibration function, and editing tools.

Technical specifications

DESR-7700

- Hard disk capacity - 250 GB
- CPU and GPU - 90 nm EE+GS
- Recordable media - DVD-R (video mode), DVD-RW (video mode, VR mode)
- Reproducible media - DVD-VIDEO, DVD-R (video mode), DVD-RW (video mode, VR mode), Music CD, CD-R (JPEG), Memory stick, "PlayStation" standard CD-ROM, "→ PlayStation 2" standard CD-ROM / DVD-ROM
- Length of video that can be recorded on a 4.7 GB DVD-R/RW:
  - HQ about 1 hour
  - HSP about 1.5 hours
  - SP (standard) about 2 hours
  - LP about 3 hours
  - EP about 4 hours
  - SLP about 6 hours
- Length of video that can be recorded on the hard drive:
  - HQ about 53 hours
  - HSP about 81 hours
  - SP (standard) About 107 hours
  - LP about 164 hours
  - EP about 217 hours
  - SLP about 325 hours
- Image record system - MPEG2
- Voice record system - Wave (WAVE files) of Linear PCM (at the time of HQ mode)
- Dolby Digital 2 Channel (at the time of HSP, SP, LP, EP, and SLP mode)
- Receiving channel - Ground analog (VHF Low:0-4CH / VHF Med:4-6CH / VHF High:7-13CH /UHF:14-83CH / CATV:CT-7-C99999CH), BS analog (1,3,5,7,9,11,13,15CH)
- Dubbing function - HDD->DVD high-speed Dubbing (12x max)
- Reproduction function - In-variable-speed reproduction
- Simultaneous Record and Playback
- Edit function A-B elimination (GOP)
  - GOP: Group of pictures
- Recording function - Electronic program table (EPG)
- High definition and sound quality 3-dimensional Y/C separation
  - Time Base Correction (TBC)
  - Video D/A converter (12 bits 108 MHz)
  - Ghost reduction tuner, DNR, Variable Bit Rate (VBR) record, Audio D/A converter (24 bits 96 kHz)
- Photograph function - File format: JPEG (DCF standard)

- Memory Stick (only Sony digital still cameras)
- USB (only Sony digital still cameras)
- Slide show display
- Picture rotation (90 degrees, 180 degrees, 270 degrees)
- Music function - File format: ATRAC3, Media: Music CD
- Network function - Firmware upgrade
- PS2 correspondence network service
- Antenna input Ground analog: VHF/UHF 75ohmF type connector
    - BSIF: 75ohmF type connector
- Input-and-output terminal D terminal output (D1/D2) x1
    - A composite image / S image / stereo sound Output terminal x1
    - A composite image / S image / stereo sound Input terminal x1
- Optical digital voice output (SPDIF) x1
- Ethernet 100 base/TX x1
- USB (Ver.1.1) terminal x1
- "Memory Stick" slot x1
- "Memory Card" slot x2
- Controller terminal x2 (→ DualShock)
- Size: 312 x 88 x 323 mm (12.5" x 3.5" x 12.75")
- Mass: About 5.8 kg (13.75 lb)

DESR-5700

- Hard disk capacity - 160 GB

DESR-7000

- Hard disk capacity - 250 GB

DESR-7100

- Hard disk capacity - 250 GB

DESR-7500

- Hard disk capacity - 450 GB

External links

- Official PSX Website (Japan) [6]
- Press Release [7]

References

[1] http://www.psx.sony.co.jp/product/DESR-7700_5700/psp.html
[2] https://cgi.sonydrive.jp/form/enquete/psx_0005/index.html
[3] http://www.psx.sony.co.jp/product/DESR-7700_5700/spec.html
[4] http://www.psx.sony.co.jp/product/DESR-7700_5700/acc.html
[5] http://www.psx.sony.co.jp/product/DESR-7700_5700/parts.html
[6] http://www.psx.sony.co.jp/
[7] http://www.sony.net/SonyInfo/News/Press_Archive/200310/03-1007E/

DualShock

The **DualShock** (Trademarked as **DUALSHOCK** and occasionally referred to as **Dual Shock**) is a line of vibration-feedback gamepads by Sony for the PlayStation, → PlayStation 2, and PlayStation 3[1] video game consoles. The DualShock was introduced in Japan in late 1997, and launched in America in May 1998, meeting with critical success. First introduced as a secondary peripheral for the original PlayStation, a revised PlayStation version came with the controller and subsequently phased out the digital controller that was originally included with the hardware, as well as the Sony Dual Analog Controller.

PlayStation 3's DualShock 3 wireless controller, which includes both vibration function and the motion-sensing functionality of the original Sixaxis wireless controller.

Models

DualShock

The **DualShock Analog Controller** (SCPH-1200) is a controller capable of providing vibration feedback based on the onscreen action of the game (if the game supports it), as well as analog input through two sticks. The controller is called "Dual*Shock*" because the controller employs two vibration motors: a weak buzzing motor that feels like cell phone or pager vibration and a strong rumble motor similar to that of the Nintendo 64's Rumble Pak. The DualShock differs from the Rumble Pak in that the Rumble Pak uses batteries to power the vibration function while all corded varieties of the DualShock use power supplied by the PlayStation. Some third party DualShock-compatible controllers use batteries instead of the PlayStation's power supply. The rumble feature of the DualShock is similar to the one featured on the first edition of the Japanese Dual Analog Controller, a feature that was removed shortly after that controller was released.

Like its predecessor, the Dual Analog controller, the DualShock controller has two analog sticks. Unlike its predecessor, however, the DualShock controller's analog sticks feature textured rubber grips instead of the smooth plastic tips with recessed grooves found on the Dual Analog controller's analog sticks. There are also additional buttons, L3 and R3, which are pressed by clicking the analogue sticks down.

The controller was hugely supported; shortly after its launch most new titles, including *Crash Bandicoot: Warped*, *Spyro the Dragon*, and *Tekken 3* included support for the vibration feature and dual analog sticks. Some games designed for the original vibration ability of the Dual Analog, such as *Porsche Challenge*, also work. Many games took advantage of the presence of two motors to provide vibration effects in stereo including *Gran Turismo* and the PlayStation port of *Quake II*. Released in 1999, the PlayStation hit *Ape Escape* became the first game to explicitly require DualShock/Dual-Analog-type controllers, with its gameplay requiring the use of both analog sticks.

DualShock 2

When the → PlayStation 2 computer entertainment system was announced, the **DualShock 2 Analog Controller** (SCPH-10010) included with it was almost exactly the same externally as the previous DualShock analog controller, except that it was black (colors came later), had different screw positioning (one fewer), and the DualShock 2 logo was added. Another way to tell the DualShock and the DualShock 2 controllers apart is that the connector that plugs into the console matches that console's memory card shape; the DualShock's connector has rounded shoulders and DualShock 2's is squared off. The analogue sticks were also noticeably stiffer for more accurate movements. Internally, the DualShock 2 was lighter and all of the buttons (except for the Analog mode, L3 and R3 buttons) were readable as analog values (pressure sensitive)[2] .

DualShock 3

Announced at the 2007 Tokyo Game Show, the **DualShock 3 Wireless Controller (SCPH-98050)** is a controller for the PlayStation 3 that incorporates all the features of the Sixaxis wireless controller with rumble capabilities. The Immersion v. Sony lawsuit has been speculated as a factor for why the Sixaxis did not have rumble capabilities.

The DualShock 3 controller was released in;

- Japan on November 11, 2007 in black at a retail price of JP¥5,500.
- North America on April 5, 2008[3] for a retail price of US$54.99.
- Australia on April 24, 2008 for a retail price of AU$99.95.
- New Zealand on May 9, 2008 for a retail price of NZ$109.95.
- Europe on July 2, 2008[4] for a retail price of €59.99.
- United Kingdom[5] on July 4, 2008 for a retail price of £39.99.

The DualShock 3 generated sales of over $10.9 million in April 2008 according to Sony Computer Entertainment America.[6] It is also bundled with the *Metal Gear Solid 4* themed 80 GB PlayStation 3, which was released on June 12 2008.[7] Hands-on accounts at the 2007 Tokyo Game Show described the controller as being capable of vibration forces comparable to the DualShock 2. According to GameSpot, DualShock 3's "rumble was a touch weak but stuck close to PlayStation 2's force feedback";[8] while various others reported more refined vibration effects than the DualShock 2, particularly with the *Metal Gear Solid 4* demonstration.[9]

The DualShock 3 is identifiable by the top labeling which incorporates both "DualShock 3" and "Sixaxis" markings. It is also easily noticeable when lifting the controller as the DualShock 3 at 192.0g weighs 40% more than the Sixaxis's 137.1g. The back markings indicate the DualShock 3 draws up to 300mA of current at 3.7V for a power consumption of 1.11 Watts, an order of magnitude increase from the 30mA of current at 3.7V (0.111 Watts) listed on the Sixaxis.

A Sony representative confirmed on 2 April 2008 that the Sixaxis controller would officially be discontinued with the release of the force-feedback enabled DualShock 3 in mid-April 2008. The Sixaxis is no longer being produced and is no longer in stock in most stores.[10]

Software requirements

PlayStation 3 firmware 1.94 or higher is required to use the DualShock 3 in compatible PlayStation 3 format software. Firmware 2.00 or higher is required to use the DualShock 3 in compatible PlayStation and PlayStation 2 format software. The first software content release supporting the DualShock 3 was the *Gran Turismo 5 Prologue* free demo made available in the Japanese PlayStation Store on October 20, 2007. A partial list of software that includes rumble support including patches (downloadable add-ons from the PlayStation Store to add rumble to software released before September 2007) was announced by SCEI at the TGS 2007.[11] Support was added to MotorStorm with an online version 3.0 patch in October, 2007.[12] In consoles with backwards compatibility, the DualShock 3 controller vibration function can be used in appropriate PS2 and PS1 titles. Future releases of games that support DualShock 3 capability will be labeled with an icon of the controller and "DualShock 3 Compatible".

Emmy Award

The DualShock controller was given an Emmy Award for "Peripheral Development and Technological Impact of Video Game Controllers" by The National Academy of Television Arts & Sciences on January 8, 2007.[13] Sony initially reported that the Sixaxis had received this award[14] before issuing a correction.[15]

References

[1] Morell, Chris (20 March 2008). " *DualShock 3 for PS3* (http://www.gamepro.com/sony/ps3/games/features/170808.shtml)". GamePro. . Retrieved 2008-03-21.

[2] " Dual Shock 2 Review (http://gear.ign.com/articles/306/306387p1.html)". IGN. 2001-09-27. . "The biggest difference between the Dual Shock 2 and the original ... is the fact that ... all of the buttons and even the digital pad offer analog support. This means that the d-pad, the four face buttons and the four shift buttons are all pressure sensitive and have 255 degrees of sensitivity. ... It's also worth noting that the Dual Shock 2 is a bit lighter than the original Dual Shock."

[3] YouTube - PS3 VIDEO NEW (DEFENSE) (http://www.youtube.com/watch?v=xHdozBNxPdQ)

[4] " DUALSHOCK3 Wireless Controller available for PLAYSTATION3 this summer (http://www.scee.presscentre.com/Content/Detail.asp?ReleaseID=4585&NewsAreaID=2)". Sony Computer Entertainment Europe. 2008-06-30. . Retrieved 2008-06-30.

[5] DualShock 3 Gets July 2 European Release | Edge Online (http://www.next-gen.biz/index.php?option=com_content&task=view&id=11147)

[6] Video Game News, Video Game Coverage, Video Game Updates, PC Game News, PC Game Coverage - GameDaily (http://www.gamedaily.com/articles/news/sony-dualshock-3-rakes-in-109-million-in-april)

[7] PlayStation.Blog » Destination PlayStation News: MGS PS3 Bundle, Kratos PSP and DualShock3 Release Date (http://blog.us.playstation.com/2008/02/26/destination-playstation-news-mgs-ps3-bundle-kratos-psp-and-dualshock3-release-date-by/)

[8] " TGS '07: Spot On - The Dual Shock 3 (http://www.gamespot.com/news/6179170.html)". *GameSpot*. CNET. 2007-09-17. . Retrieved 2008-01-11.

[9] Pigna, Kris (2007-09-26). " Kojima Productions Says DualShock 3 Better DS2 (http://www.1up.com/do/newsStory?cId=3163213)". *1UP.com*. Ziff Davis. . Retrieved 2008-01-11.

[10] MTV Multiplayer » Sony Non-Shocker: Sixaxis Discontinued (http://multiplayerblog.mtv.com/2008/04/02/sony-non-shocker-sixaxis-discontinued/)

[11] " PlayStation 3 TGS2007 official site DualShock3 announcement (http://www.jp.playstation.com/tgs2007/ps3/hardware/index.html)" (in Japanese). 2007. . Retrieved 2008-01-11.

[12] McWhertor, Michael (2007-10-24). " Motorstorm 3.0 patch adds DualShock 3 fun (http://kotaku.com/gaming/patch-watch/motorstorm-30-patch-coming-adds-dualshock-3-fun-314677.php)". . Retrieved 2008-01-11.

[13] " National Television Academy Announces Emmy Winning Achievements: Honors Bestowed at 58th Annual Technology & Engineering Emmy Awards (http://www.emmyonline.org/mediacenter/tech_2k6_winners.html)". 2007-09-20. . Retrieved 2008-01-11.

[14] " Sony Computer Entertainment America Wins Emmy Award for PLAYSTATION3 SIXAXIS Wireless Controller (http://web.archive.org/web/20070110162026/http://www.us.playstation.com/News/PressReleases/374)". Sony Computer Entertainment America. 2007-01-08. Archived from the original (http://www.us.playstation.com/News/PressReleases/374) on 2007-01-10. . Retrieved 2008-01-11.

[15] Thorsen, Tor (2007-01-10). " Sony retracts Sixaxis Emmy claims (http://www.gamespot.com/news/6164037.html)". *GameSpot*. CNET. . Retrieved 2008-01-11.

PlayStation 2 Expansion Bay

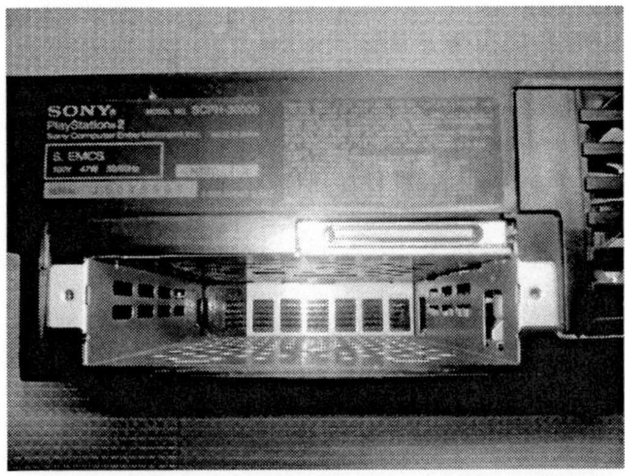

PlayStation 2 Expansion Bay on SCPH-30000

The **PlayStation 2 Expansion Bay** is a 3.5" drive bay introduced with the model 30000 and 50000 → PlayStation 2 (replacing the PCMCIA slot used in the models 10000, 15000, and 18000, and no longer present as of the model 70000) designed for the network adaptor and internal hard disk drive (HDD). These peripherals enhance the capabilities of the PS2 to allow online play and other features and were shown at E³ 2001.

Network adaptor

See also: List of PlayStation 2 network games

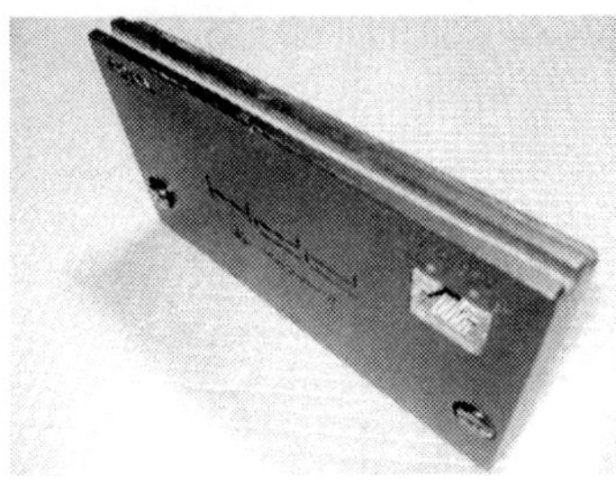
Network Adapter for PlayStation 2

The network adaptor was released on July 19, 2001 in Japan (together with the Hard disk drive) , in August 2002 in North America and in June 2003 in Europe. Two models were available - one with a dial-up modem and an Ethernet (RJ-45) jack for broadband Internet connection (sold in North America), and one with only an Ethernet interface (sold in Europe and other regions). A start-up disc ("Network Access Disc") is included with the Network Adaptor and installs a file on the memory card for connection settings which are accessible by all but one Network Adaptor compatible game. *Tony Hawk's Pro Skater 3* was released in November 2001 and supported the Network Adaptor hardware, but not the software as it was not finalised until much later.

The Network adaptor also provides a Parallel ATA interface and a Molex disk drive power connector to allow installation of a 3.5" IDE hard disk drive in the expansion bay.

The Slim-line range of the PS2 has an Ethernet jack built-in, but no hard disk drive interface.

Hard disk drive

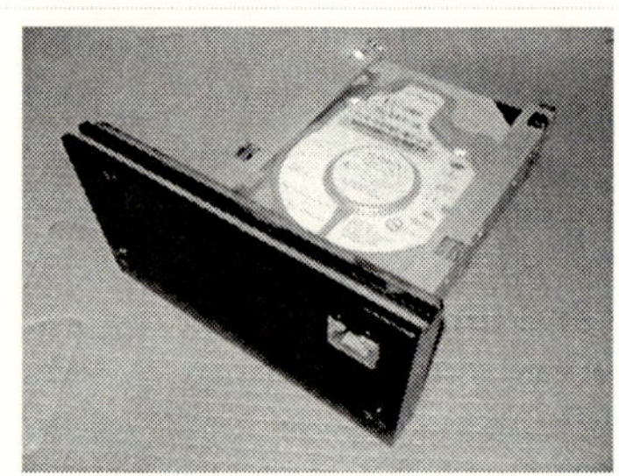
PlayStation BB Unit, HDD and Network Adapter

The **PlayStation 2 Hard Disk Drive** (PS2 HDD) was released on July 19, 2001 in Japan (together with the Network Adaptor) and on March 23, 2004 in North America. It requires the Network Adaptor to connect to the → PlayStation 2 and to receive power. The HDD has a 40 GB capacity that can be used by games to reduce load time by putting data on the hard drive temporarily, or back up memory card data. Due to MagicGate copyright protection, programs that are bootable directly from the HDD (ex. PlayStation Broadband Navigator, PlayOnline Viewer, Pop'n Music Puzzle-dama Online) are keyed to the system when that system installs them. The HDD can be transferred to another PlayStation 2 system and files on the HDD can be accessed, but those specific programs cannot be booted without being reinstalled. Contrary to popular belief, a complete reformat of the HDD is not necessary upon transfer of the HDD between consoles, or else it would not be useful to have the HDD be preformatted and have preinstalled software, as is the case with the North American HDD unit. A HDD Utility Disc is included to allow maintenance of the HDD and in North America, *Final Fantasy XI* is also included. Currently, there are 29 North American games that support the HDD.

Unofficial software called HD Loader and the newer HD Advance allow users to copy entire games to the HDD and run them without the discs. These pieces of software also allow you to use some standard HDDs in the PS2, however they will not be seen in the browser, or work with games using the HDD. While some argue that this improves performance and protects the potentially fragile discs, especially from young children, others claim that this only encourages piracy as rented games can be copied and kept forever (see rental piracy).

As of December 2004[1]

```
, it is widely believed that Sony is no longer interested in supporting
 the hard drive. The new slimline model of PlayStation 2, which
replaced the older, larger model in November 2004, is not capable of
using the hard drive and while Sony has stated they are investigating
alternatives, it is thought that none will be forthcoming due to the
relatively late date in the product's lifecycle. The absence of support
 for the HDD in the slimline PS2 also means that Final Fantasy XI can
only be played on older PlayStation 2s. Sony's public response at the
time of the introduction of the slimline PS2 (in mid-2004) was that
"the more hardcore [HDD-interested] gamers...already have their
PlayStation 2 units," according to a company statement.
```

North American Releases with HDD Support

- *Final Fantasy XI* is the only North American game truly dependent on the HDD as it requires various patches and upgrades from Square Enix.
- *SOCOM 3: U.S. Navy SEALs* supports additional maps, downloadable via the in game "Socom Store" used to be $5.99 until the third pack was released late March 2008; now all three are free.
- *SOCOM II: U.S. Navy SEALs* supports additional maps, however the files must be copied from a magazine demo disc to the hard drive and cannot be downloaded.
- *SOCOM: U.S. Navy SEALs Combined Assault* supports additional maps.
- *Resident Evil: Outbreak* (Both File #1 and File #2) installs 1 GB to the HDD for reduced loading times.
- *ESPN NFL 2K5*, *ESPN NBA 2K5*, *ESPN College Hoops 2K5*, *ESPN MLB 2K5*, and *ESPN NHL 2K5* use the HDD to improve replays. (If the HDD is not installed, static screenshots are shown as replays. With the HDD, full cutscene-like replays can be displayed.). ESPN NHL 2K5 has the ability to save files directly to the hard drive.
- *RPG Maker 3* installs 3 GB to the HDD to decrease load time.
- 2K Sports's *NBA 2K6*, *MLB 2K6*, *NHL 2K6*, *College Hoops 2K6*, and their respected updates; up to the 2K9 versions (only until 2K8 for College Hoops), also use the HDD to display recorded replays from game action. Without it, stills are shown in *NBA 2K6* (during halftime and the end of the game) and no end-of-inning replays are shown in *Major League Baseball 2K6*.
- *Metal Saga* installs 1,7GB to the HDD to decrease load time and uses the HDD to save/load game instead of Memory Card
- *Street Fighter Alpha Anthology*, like its Japanese counterpart, can install 2 GB to the HDD to reduce loading time.
- *The Urbz: Sims in the City* recognizes when the HDD is installed and allows data to be saved directly to it. Also creates a 512mb "_tmp" (if not already there) Partition to cache Game Files to speed up load times.
- *The Sims 2* recognizes when the HDD is installed and allows data to be saved directly to it. Also creates a 512mb "_tmp" (if not already there) Partition to cache Game Files to speed up load times.

Japanese Releases with HDD Support

- *Final Fantasy X* (and *Final Fantasy X International*) installs a 1,664 megabyte file to the HDD to reduce load times
- *Kingdom Hearts* (and *Kingdom Hearts Final Mix*) installs a 1,280 megabyte file to the HDD to reduce load times.
- *Xenosaga Episode 1* installs a 1,792 megabyte file to the HDD to reduce load times. It allows the game to be saved to and loaded from the HDD instead of a Memory Card.
- *PlayOnline Viewer* fully installs to the HDD so that it can be patched/updated, currently using 1 gigabyte. It is used to boot *Final Fantasy XI*, *Front Mission Online*, *Tetra Master*, and *JongHowLo*.
- *Tetra Master* fully installs to the HDD so that it can be patched/updated, currently using 128 megabytes. It comes with *Final Fantasy XI*.
- *JongHowLo* fully installs to the HDD so that it can be patched/updated, currently using 256 megabytes. It comes with *Final Fantasy XI*.
- *Final Fantasy XI* (and the *Rise of the Zilart*, *Chains of Promathia*, *Treasures of Aht Urhgan*, and *Wings of the Goddess* expansions) fully installs to the HDD so that it can be patched/updated. Uses 8,192 megabytes, with both expansions installed, as of March 7th, 2005 (Note: This does not include the Treasures of Aht Urgahn and Wings of the Goddess expansions as they were released after).
- *Unlimited Saga* installs a 3,072 megabyte file to the HDD to reduce load times.
- *Energy Airforce* Allows playback of music that is stored on the HDD while playing the game instead of listening to the music that is built into the game. (requires PlayStation Broadband Navigator 0.20 or higher)
- *3D Fighting School 2* allows the game to be saved to and loaded from the HDD instead of a Memory Card.
- *Dark Chronicle* (aka *Dark Cloud 2*) installs a 1,536 megabyte file to the HDD to reduce load times. The Asia version also supports the HDD (most Asia versions of games have HDD support removed).]

- *Bomberman Kart*, has downloadable content with new tracks to install on HDD.
- *Capcom vs. SNK 2* installs to the HDD to reduce load times and allows the game to be saved to and loaded from the HDD instead of a Memory Card.
- *Zettai Zetsumai Toshi* (aka *Disaster Report*) installs to the HDD to reduce load times and allows the game to be saved to and loaded from the HDD instead of a Memory Card.
- *Soul Calibur II* installs to the HDD to reduce load times.
- *Star Ocean: Till the End of Time* (and the Director's Cut) installs to the HDD to reduce load times. Installation also reduces occurrences of a game crashing glitch that is known to happen on the first batch of discs when played on model 1x000 PS2s.
- *Pop'n Taisen Puzzle-dama Online* Installation to the HDD is required to play. The game boots from PSBBN or HDD Utility Disc and does not require the disc or a registration code, making it a very unusual case of HDD support, as it has no anti-piracy protection to prevent the disc from being passed around in a group of people.
- *DJ Box* Sony Computer Entertainment's MP3 DJ mixing program requires the hard drive for MP3 storage. Also you can save your own DJ mixes that you make with the game to the hard drive.
- *Dirge of Cerberus: Final Fantasy VII* requires the hard drive for online play. The game uses Square-Enix's PlayOnline service, which needs the hard drive so you can install the online interface. Square-Enix also plans to patch the game with the aid of the hard drive.
- *Romancing SaGa Minstrel's Song* installs a 5,120 megabyte file to the HDD to reduce load times.
- *Shin Sangoku Musou 3* installs a 512 megabyte file to the HDD to reduce load times.
- *Shin Sangoku Musou 3 Moushouden* installs a 1 gigabyte file to the HDD to reduce load times.
- *Street Fighter ZERO Fighter's Generation* installs a 2 gigabyte file to the HDD to reduce loading time.
- *Guitar Freaks 4th Mix & DrumMania 3rd Mix* installs to the HDD to reduce load times.
- *Beatmania IIDX 5th Style* can cache song files to the HDD to reduce load times. In addition, using HDD caching will enable a bonus gameplay mode.
- *Age of Empires II: The Age of Kings* installs 128mb to fast up the game, so it no longer "Freezes" in game to load the needed data from the Game CD-ROM.
- *Musou Orochi Maou Sairin* installs 1GB to speed up load times.
- *Winning Eleven 6: Final Evolution* installs 512mb to speed up load times.
- *Winning Eleven 8: Liveware Evolution* installs 1GB to speed up load times.
- *Gundam Musou Special* installs 512mb to speed up load times.
- *Gundam Musou 2* installs 512mb to speed up load times
- *Shin Sangoku Musou 5 Special* installs 512mb to speed up load times.
- *Let's Bravo Music* Let you download extra Music/Adventures to the Hdd.
- *G1 Jockey 4 2007 & Winning Post 7 2007* Installs each 1GB to speedup load times.
- *TVware* the whole TVware series uses the hdd
- *A Ressha de Gyoukou 2001 Perfect Set* (A-Train 2001 + Train-Pack) installs 256mb to the hdd, and the Train-Pack installs additional 14mb.
- *Ace Combat 4 installs 1gb to speed up load times.*
- *The Urbz: Sims in the City* recognizes when the HDD is installed and allows data to be saved directly to it. Also creates a 512mb "_tmp" (if not already there) Partition to cache Game Files to speed up load times.
- *The Sims 2* recognizes when the HDD is installed and allows data to be saved directly to it. Also creates a 512mb "_tmp" (if not already there) Partition to cache Game Files to speed up load times.
- *Metal Saga* installs 1,7GB to the HDD to decrease load time and uses the HDD to save/load game instead of Memory Card
- *Biohazard: Outbreak* (Both File #1 and File #2) installs 1 GB to the HDD for reduced loading times.
- *G1 Jockey 4 2008* installs 1GB to speed up load times
- *Vampire: Darkstalkers Collection* installs 1GB to speed up load times.

Linux kit

Main article: → PS2 Linux

The **Linux Kit** for PlayStation 2 was released in 2002 and included the PlayStation 2 Linux software, keyboard, mouse, VGA adapter (which requires an RGB monitor with sync-on-green), Network Adaptor (Ethernet only) and a 40 GB hard disk drive. It allows the PlayStation 2 to be used as a personal computer.

External links

- Network Adaptor official site [2]
- PlayStation 2 Linux [3]

References

[1] http://en.wikipedia.org/wiki/Playstation_2_expansion_bay
[2] http://www.us.playstation.com/peripherals.aspx?id=SCPH-10281
[3] http://playstation2-linux.com/

EyeToy

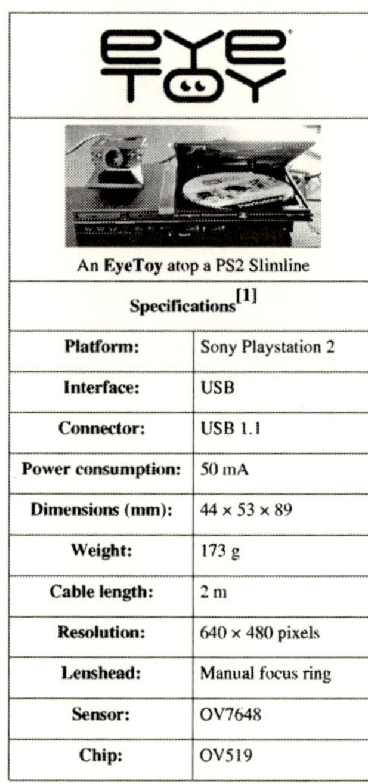

An **EyeToy** atop a PS2 Slimline

Specifications[1]	
Platform:	Sony Playstation 2
Interface:	USB
Connector:	USB 1.1
Power consumption:	50 mA
Dimensions (mm):	44 × 53 × 89
Weight:	173 g
Cable length:	2 m
Resolution:	640 × 480 pixels
Lenshead:	Manual focus ring
Sensor:	OV7648
Chip:	OV519

The **EyeToy** is a color digital camera device, similar to a webcam, for the → PlayStation 2. The technology uses computer vision and Gesture recognition to process images taken by the camera. This allows players to interact with games using motion, color detection and also sound, through its built-in microphone.

The camera is manufactured by → Logitech (known as "Logicool" in Japan), although newer EyeToys are manufactured by Namtai. The camera is mainly used for playing EyeToy games developed by Sony and other companies. It is not intended for use as a normal PC camera, although some people have developed unofficial drivers for it.[2] As of November 6, 2008, the EyeToy has sold 10.5 million units worldwide.[3]

History

The EyeToy was conceived by Richard Marks in 1999, after witnessing a demonstration of the PlayStation 2 at the 1999 Game Developers Conference in San Jose, California.[4] Marks's idea was to enable natural user interface and mixed reality video game applications using an inexpensive webcam, using the computational power of the PlayStation 2 to implement computer vision and gesture recognition technologies. He joined Sony Computer Entertainment America (SCEA) that year, and worked on the technology as Special Projects Manager for Research and Development.[5] [6]

Marks's work drew the attention of Phil Harrison, then Vice President of Third Party Relations and Research and Development at SCEA. Soon after being promoted to Senior Vice President of Product Development at Sony Computer Entertainment Europe (SCEE) in 2000, Harrison brought Marks to the division's headquarters in London to demonstrate the technology to a number of developers. At the demonstration, Marks was joined with Ron Festejo of SCE Camden Studio[6] (which would later merge to become SCE London Studio) to begin developing a software title using the technology, which would later become *EyeToy: Play*. Originally called the iToy (short for "interactive toy") by the London branch, the webcam was later renamed to the EyeToy by Harrison. It was first demonstrated to the public at the PlayStation Experience event in August 2002 with four minigames.[5]

Already planned for release in Europe, the EyeToy was picked by SCE's Japanese and American branches after the successful showing at the PlayStation Experience. In 2003, EyeToy was released in a bundle with *EyeToy: Play*: in Europe on July 4, and North America on November 4. By the end of the year, the EyeToy sold over 2 million units in Europe and 400,000 units in the United States.[5] On February 11, 2004, the EyeToy was released in Japan.

Design

The camera is mounted on a pivot, allowing for positioning. Focusing the camera is performed by rotating a ring around the lens. It comes with two LED lights on the front. A blue light turns on when the PS2 is on, indicating that it is ready to be used, while the red light flashes when there is insufficient light in the room. There is also a microphone built in. A second, newer model of the EyeToy provides similar features, but sports a smaller size and silver casing.[7]

On computer

Because the EyeToy is essentially a web camera inside a casing designed to match the Sony PlayStation 2, and it uses a USB 1.1 protocol and USB plug, drivers have been created to make it work with many operating systems. The type of driver required depends on the model of EyeToy camera. There are three different types:

- SLEH-00031
- SCEH-0004
- SLEH-00030

The model information is included in a label on the bottom of the camera.

In addition, the red LED that normally signals inadequate lighting is used as the active recording indicator. The blue LED is lit when the EyeToy is plugged into the computer

Technical limitations

Due to the camera's need to "see" the player as they play, the camera needs to be used in a well-lit room. To help let the player know when there is not enough light, there is a red LED on the front of the camera that flashes when it is too dark. Some of the games utilise this feature, requiring players to place their hand in front of the camera to quit a certain mode.Wikipedia:Citation needed

In response to this limitation, Sony has filed a patent for a "wand" controller capable of illuminating different colored LEDs in order to communicate the controller's position and simple commands to the camera.[8]

Games

Designed for EyeToy

These games require the EyeToy to be played. *All produced by Sony unless noted.*

- 2003
  - *EyeToy: Play*
  - *EyeToy: Groove*
- 2004
  - *Disney Move* (Ubisoft)
  - *EyeToy: Antigrav*
  - *EyeToy: Monkey Mania*
  - *EyeToy: Play 2*
  - *Nicktoons Movin'* (THQ)
  - *Sega SuperStars* (Sega)
  - *U Move Super Sports* (Konami)
- 2005
  - *Clumsy Shumsy* (Phoenix Games Ltd. [UK/NL])
  - *EyeToy: Chat* - a videophone system for use with the network adaptor
  - *EyeToy: EduKids*
  - *EyeToy: Kinetic*
  - *EyeToy: Operation Spy* (known as SpyToy in Europe)
  - *EyeToy: Play 3*
- 2006
  - *Eyetoy: Kinetic Combat*
  - *Eyetoy: Play Sports*
  - *Rhythmic Star* (Namco)
- 2007
  - *EyeToy: Astro Zoo*
- 2008
  - *EyeToy Play: Hero*
  - *EyeToy Play: PomPom Party*
- Unreleased
  - *EyeToy: Fight*
  - *EyeToy: Tales*

Optional EyeToy features

These games may be used with the EyeToy optionally. They have an "EyeToy Enhanced" label on the box.

- *AFL Premiership 2005* (Sony, 2005)
- *Buzz! The Music Quiz* (Sony, late 2005)
- *Buzz! The Big Quiz* (Sony, March 2006)
- *Dance Dance Revolution Extreme (North America)* (Konami, 2004) - EyeToy mini games, players can optionally see themselves dancing, additional mode with 2 camera targets.
- *DDR Festival Dance Dance Revolution* (Konami, 2004) - EyeToy mini games, players can optionally see themselves dancing, additional mode with 2 camera targets.

- *Dancing Stage Fusion* (Konami, 2004) - EyeToy mini games, players can optionally see themselves dancing, additional mode with 2 camera targets.
- *Dance Dance Revolution Extreme 2* (Konami, 2005) - EyeToy mini games, players can optionally see themselves dancing, additional mode with 2 camera targets.
- *Dancing Stage Max* (Konami, 2005) - EyeToy mini games, players can optionally see themselves dancing, additional mode with 2 camera targets.
- *Dance Dance Revolution Strike* (Konami, 2006) - EyeToy mini games, players can optionally see themselves dancing, additional mode with 2 camera targets.
- *Dance Dance Revolution SuperNova (North America)* (Konami, 2006) - EyeToy mini games, players can optionally see themselves dancing, additional mode with 2 camera targets.
- *Dance Dance Revolution SuperNova* (Konami, 2007) - EyeToy mini games, players can optionally see themselves dancing, additional mode with 2 camera targets.
- *Dancing Stage SuperNova (Europe)* (Konami, 2007) - EyeToy mini games, players can optionally see themselves dancing, additional mode with 2 camera targets.
- *Dance Dance Revolution SuperNova 2 (North America)* (Konami, 2007) - EyeToy mini games, players can optionally see themselves dancing, additional mode with 2 camera targets.
- *Dance Dance Revolution SuperNova 2* (Konami, 2008) - EyeToy mini games, players can optionally see themselves dancing, additional mode with 2 camera targets.
- *Dance Dance Revolution X (North America)* (Konami, 2008) - EyeToy mini games, players can optionally see themselves dancing, additional mode with 2 camera targets.
- *Dance Factory* - players can optionally see themselves dancing, additional mode with 2 camera targets.
- *DT Racer* (XS Games, 2005) - photo taken by EyeToy can be used as a custom avatar
- *Formula One 05* (Sony, mid 2004)
- *Flow: Urban Dance Uprising*
- *Get On Da Mic* (Eidos, 2004) - players can see their performance
- *Harry Potter and the Prisoner of Azkaban* (EA, 2004) - features EyeToy minigames
- *Jackie Chan Adventures* (Sony, 2004) - features Eye Toy minigames
- *Lemmings* (Team 17, 2006)
- *LMA Manager 2005* (Codemasters, 2004) - players can have their pictures on in-game newspapers
- *NBA 07*
- *Racing Battle: C1 Grand Prix* (Genki, 2005) - Used to capture textures to be used as car stickers in the bodypaint interface[9]
- *SingStar* series (Sony, 2004-2008) - singers can optionally see themselves when singing
- *The Sims 2*
- *Stuart Little 3: Big Photo Adventure*
- *The Polar Express* (THQ, 2004)
- *The Sims 2: Pets*
- *The Urbz: Sims in the City* (EA, 2004) - players can have their faces on in-game billboards
- *Tony Hawk's Underground* (Activision/Neversoft, 2003) -Player can capture an image of their face and map it onto their character.
- *YetiSports Arctic Adventures* (JoWooD, 2005) - EyeToy multi-player games
- *Who Wants To Be A Millionaire? Party Edition* (Eidos, late 2006) - players can have their 'mugshots' on a winning check
- *World Tour Soccer 2006*

Cameo

EyeToy: Cameo is a system for allowing players to include their own images as avatars in other games. Games that support the feature include a head scanning program that can be used to generate a 3D model of the player's head. Once stored on a memory card, this file is then available in games that support the Cameo feature. *EyeToy: Cameo* licenses the head creation technology Digimask.

Supported games

- *AFL Premiership 2005*
- *AFL Premiership 2006*
- *AND 1 Streetball*
- *CMT Presents: Karaoke Revolution Country*
- *EyeToy: Kinetic*
- *EyeToy: Play*
- *EyeToy: Play 2*
- *EyeToy: Play 3*
- *Formula One 05*
- *Gaelic Games: Football*
- *Gretzky NHL 2005*
- *Karaoke Revolution Party*
- *Karaoke Revolution Presents: American Idol*
- *MLB 2005*
- *MLB '06: The Show*
- *MLB '07: The Show*
- *MLB '08: The Show*
- *Sims 2*
- *The Urbz*
- *This Is Football 2005*
- *Tony Hawk's Underground 2*
- *Tony Hawk's American Wasteland*
- *World Tour Soccer 2006*

See also

- List of EyeToy games
- Augmented virtuality
- Dreameye - The very first camera accessory for a home gaming console, used on the Sega Dreamcast
- PlayStation Eye - The successor to the EyeToy for the PlayStation 3
- Xbox Live Vision - A similar camera made for the Xbox 360

External links

- EyeToy official site [10]
- EyetoyOnComputer Project [11] Free Automatically Installing Drivers and capture programs to make the EyeToy work on Mac OS X, Windows and Linux computers. This project has not yet released any files for Linux, use the below link for Linux drivers.
- ov51x-jpeg [12] The ov51x driver for Linux, modified to support the EyeToy.
- EyeToy webcam drivers for XP [13] Free drivers and tutorial for using the EyeToy as a webcam for Windows XP.
- EyeToy webcam drivers, tutorial and FAQ for Windows 98SE, Millennium, Windows 2000, XP 32Bit and Vista 32Bit [14].

References

[1] EyeToy specifications, published by Sony with EyeToy instruction manual.
[2] http://en.wikipedia.org/wiki/Eyetoy#endnote_drivers
[3] Tom Kim (2008-11-06). " In-Depth: Eye To Eye - The History Of EyeToy (http://www.gamasutra.com/php-bin/news_index.php?story=20975)". Gamasutra. . Retrieved 2008-11-15.
[4] Robischon, Noah (2003), " Smile, Gamers: You're in the Picture (http://www.nytimes.com/2003/11/13/technology/smile-gamers-you-re-in-the-picture.html)", *The New York Times* (The New York Times Company): G1, 2003-11-13, ISSN 0362-4331 (http://worldcat.org/issn/0362-4331), OCLC 1645522 (http://worldcat.org/oclc/1645522), , retrieved 2009-06-10
[5] Pham, Alex (2004), " EyeToy Springs From One Man's Vision (http://articles.latimes.com/2004/jan/18/business/fi-eyetoy18)", *Los Angeles Times*: C1, 2004-01-18, ISSN 0458-3035 (http://worldcat.org/issn/0458-3035), OCLC 3638237 (http://worldcat.org/oclc/3638237), , retrieved 2009-06-10
[6] Richard Marks. (2004-01-21) (Windows Media v7). *EyeToy: A New Interface for Interactive Entertainment* (http://lang.stanford.edu/courses/ee380/2003-2004/040121-ee380-100.wmv). Stanford University. Event occurs at 08:22. . Retrieved 2009-06-20.
[7] Drivers for Windows and Linux free and with support from: http://eyetoy8057.sourceforge.net/cms/
[8] Invention: Magic wand for gamers - info-tech - 23 August 2005 - New Scientist (http://www.newscientist.com/article.ns?id=dn7890)
[9] 充実のボディペイント機能 (http://www.genki.co.jp/games/rb/sp_2/0428_1.html)
[10] http://www.eyetoy.com
[11] http://eocp.sourceforge.net/
[12] http://www.rastageeks.org/ov51x-jpeg/index.php/Main_Page
[13] http://www.iplayplaystation.com/eyetoy-as-webcam/
[14] http://members.driverguide.com/driver/detail.php?driverid=487292

PlayStation 2 Headset

The **PlayStation 2 Headset** is a USB headset used with the → PlayStation 2. While the original headset was produced by → Logitech and distributed with *SOCOM*, other headsets that support the *usb-audio* class may be compatible.

The PlayStation 2 headset can also be used on PCs as it is a standard USB headset. No drivers are required in Windows, Mac OS X, or Linux.

The headset is most commonly used in online multiplayer games, however it can also be used in some karaoke style games, for voice control, and to enhance the immersive experience of some single player games.

Games that support the headset

- *25 To Life*
- *All Star Baseball 2005*
- *Area 51*
- *ATV Offroad Fury 3*
- *Battlefield 2: Modern Combat*
- *Champions of Norrath*
- *Champions: Return to Arms*
- *Call of Duty 2: Big Red One*
- *Call of Duty 3*
- *Destruction Derby Arenas*
- *ESPN NBA Basketball*
- *ESPN NCAA Basketball*
- *ESPN NFL Football*
- *ESPN NHL Hockey*
- *ESPN MLB Baseball*
- *FIFA Football 2004*
- *FIFA Soccer 2007*
- *Fight Night 2004*
- *Greg Hasting's Tournament Paintball Max'd*
- *Hardware: Online Arena*
- *Jak X: Combat Racing*
- *James Bond 007: Everything or Nothing*
- *Karaoke Revolution*
- *Killzone*
- *Lifeline*
- *The Lord of the Rings: The Return of the King*
- *Madden NFL 2004*
- *Manhunt* (in an ambient capacity)
- *Medal of Honor: Rising Sun*
- *Metal Gear Solid 3: Subsistence*
- *MLB 2005*
- *MLB Slugfest: Loaded*
- *MVP Baseball 2004*
- *NASCAR Thunder 2004*
- *NASCAR 2006: Total Team Control*

- *NASCAR 07*
- *NBA Live 2004*
- *NBA Shootout 2004*
- *NFL Gameday 2004*
- *NFL Street*
- *NHL 2004*
- *NCAA Final Four 2004*
- *NCAA Football 2004*
- *NCAA Gamebreaker 2004*
- *NCAA March Madness 2004*
- *Need for Speed: Underground 2*
- *Ratchet & Clank: Up Your Arsenal*
- *Ratchet: Deadlocked*
- *Resident Evil: Outbreak*
- *Risk: Global Domination*
- *Sly 2: Band of Thieves*
- *Sly 3: Honor Among Thieves*
- *SOCOM: U.S. Navy SEALs*
- *SOCOM II: U.S. Navy SEALs*
- *SOCOM 3: U.S. Navy SEALs*
- *SOCOM: U.S. Navy SEALs Combined Assault*
- *SSX 3*
- *Star Wars: Battlefront*
- *Star Wars Battlefront 2*
- *SWAT: Global Strike Team*
- *Syphon Filter: The Omega Strain*
- *This Is Football 2004*
- *TimeSplitters: Future Perfect*
- *Tiger Woods PGA Tour 2004*
- *ToCA Race Driver 3*
- *Tom Clancy's Ghost Recon*
- *Tom Clancy's Ghost Recon: Jungle Storm*
- *Tom Clancy's Rainbow Six 3*
- *Tom Clancy's Splinter Cell: Pandora Tomorrow*
- *Tom Clancy's Splinter Cell: Chaos Theory*
- *Tom Clancy's Splinter Cell: Double Agent*
- *Trivial Pursuit: Unhinged*

Logitech

Logitech	
Type	Public (SWX: **LOGN** [1]) (NASDAQ ADRs: **LOGI** [2])
Founded	1981
Headquarters	Romanel-sur-Morges, Switzerland
Key people	Guerrino De Luca (President and CEO) Mark J. Hawkins (SVP, Finance and Information Technology, and CFO)
Industry	Peripherals
Products	Peripherals
Revenue	▯ US$ 2.370 billion (2008)[3]
Net income	▯ US$ 849.12 million (2008)[3]
Total assets	▯ US$ 286.12 million (2008)[3]
Total equity	▯ US$ 231.03 million (2008)[3]
Employees	7,500
Website	www.logitech.com [4]

Logitech International S.A. (SWX: **LOGN** [1], NASDAQ: LOGI [5]), headquartered in Romanel-sur-Morges, Switzerland, is the holding company for Logitech Group, a Swiss peripheral-device maker. Logitech makes peripheral devices for PCs, including keyboards, mice, game controllers and webcams. Logitech also makes home and computer speakers, headphones, wireless audio devices, as well as audio devices for MP3 players and mobile phones.

Logitech's Silicon Valley office in Fremont

In addition to its Swiss headquarters, the company has offices in Fremont, California, as well as throughout Europe, Asia and the rest of Americas. Logitech's sales and marketing activities are organized into six geographic regions: Americas, Europe, Middle East, Africa, Australia, and Asia Pacific.

Brand names

In the Japanese market, Logitech uses the brand name Logicool since a Logitec (ロジテック *rojitekku*) that focuses on computer peripheral devices has existed in that country since 1982, and its parent company has used the Logitec [*sic*] brand name since 1974.

In the UK, Logitech trades under 'Logi (UK) Ltd'; a 'Logitech' based in Glasgow, Scotland manufactures precision cutting, lapping and polishing equipment for the materials processing industry. In Canada, Logitech International uses its own name without conflict with Logitech Electronics, an InterTAN Canada Ltd. supplier of consumer electronics since 1988.

Since the 1980s, Logitech has made computer mice and keyboards directly for Apple Inc.,HP, Dell and other companies.

History

Logitech International S.A. was co-founded in Apples, Vaud, Switzerland, in 1981 by two Stanford Masters alumni, Daniel Borel and Pierluigi Zappacosta, and Giacomo Marini, formerly a manager at Olivetti.

The mass-marketed computer mouse was the product that made Logitech well-known. The range of products offered improvements over a product originally developed at LAMI (École polytechnique fédérale de Lausanne) by professor Jean-Daniel Nicoud and engineer André Guignard, who was involved in the design changes of the computer mouse originally invented by Douglas Engelbart.

For a time during its formative years, Logitech's Silicon Valley offices occupied space at 165 University Avenue, Palo Alto, California, home to a number of noted technology startups.

From there, Logitech expanded its product line (see below) to encompass many mass market computer peripherals and beyond (such as the "Harmony" range of programmable universal remote controls).

Production

The first Logitech mice were made in Le Lieu, in the Swiss Canton of Jura by Dubois Depraz SA.

Production facilities were then established in the US, Taiwan, Ireland and moved subsequently to Suzhou, China. As of 2005[6]

```
, the manufacturing operations in China produce approximately half of
Logitech's products.  The remaining production is outsourced to
 contract manufacturers and original design manufacturers in Asia.
```

In December 2008, Logitech announced the production of its billionth mouse.[7]

Products

A Logitech game controller.

- Keyboards, computer mice, and trackballs (wired and wireless models).
- QuickCam webcams.
- PC speakers, including stereo as well as 2.1, 5.1 and 7.1 channel surround sound systems.
- Logitech 'G' series PC gaming hardware.
- Xbox, Xbox 360, → PS2, PS3 and PSP gaming hardware, including game controllers, joysticks, keyboards and racing wheels.
- Headphones, headsets and desktop microphones.
- iPod, PSP, MP3 player and mobile phone accessories. Including iPod and PSP speaker docks.
- Harmony universal remotes.
- Squeezebox wireless music systems.
- io2 Digital Writing System.
- Ultimate Ears headphones and in ear monitors.

Sponsorships

Logitech is the main club sponsor of Co Carlow FC for the 2007/2008 season.

See also

- 3Dconnexion (Subsidiary of Logitech)
- Labtec (Subsidiary of Logitech)

References

[1] http://www.swx.com/market/quote_chart_en.html?id=CH0021655334CHF4
[2] http://quotes.nasdaq.com/asp/SummaryQuote.asp?symbol=LOGI
[3] Google. " "Logitech International SA (USA) (Public, NASDAQ:LOGI)" (http://www.google.com/finance?client=ob&q=NASDAQ:LOGI)". *microsoft.com*. NASDAQ. . Retrieved Apr 9, 4:36PM EDT.
[4] http://www.logitech.com/
[5] http://quotes.nasdaq.com/asp/SummaryQuote.asp?symbol=LOGI&selected=LOGI
[6] http://en.wikipedia.org/wiki/Logitech
[7] " Firm makes one billionth mouse (http://news.bbc.co.uk/2/hi/technology/7751627.stm)". BBC News. . Retrieved 2008-12-03.

External links

- Logitech (http://www.logitech.com)

Articles

- Mouse maker reveals how niche players succeed (http://www.swissinfo.org/eng/business/detail/Mouse_maker_reveals_how_niche_players_succeed.html?siteSect=161&sid=6645666), Swissinfo, 20.4.2006

Linux for PlayStation 2

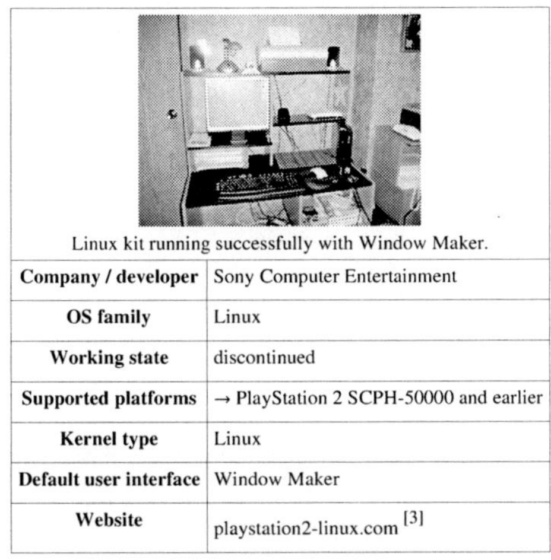
Linux kit running successfully with Window Maker.

Company / developer	Sony Computer Entertainment
OS family	Linux
Working state	discontinued
Supported platforms	→ PlayStation 2 SCPH-50000 and earlier
Kernel type	Linux
Default user interface	Window Maker
Website	playstation2-linux.com [3]

Linux for PlayStation 2 (or PS2 Linux) is a kit released by Sony Computer Entertainment in 2002 that allows the → PlayStation 2 console to be used as a personal computer. It included a Linux-based operating system, a USB keyboard and mouse, a VGA adapter, a PS2 network adaptor (Ethernet only), and a 40 GB hard disk drive (HDD). An 8 MB memory card is required; it must be formatted during installation, erasing all data previously saved on it, though afterwards the remaining space may be used for savegames. It is strongly recommended that a user of Linux for PlayStation 2 have some basic knowledge of Linux before installing and using it, due to the command-line interface for installation.

The official site for the project will be closed by the end of October 2009 [1] but communities like ps2dev [2] remains active.

Capabilities

The Linux Kit turns the PlayStation 2 into a full-fledged computer system, but it does not allow for use of the DVD-ROM drive except to read PS1 and PS2 discs due to piracy concerns by Sony.Wikipedia:Citation needed Although the HDD included with the Linux Kit is not compatible with PlayStation 2 games, reformatting the HDD with the utility disc provided with the retail HDD enables use with PlayStation 2 games but erases PS2 Linux, though there is a driver that allows PS2 Linux to operate once copied onto the APA partition created by the utility disc. The Network Adaptor included with the kit only supports Ethernet; a driver download is available to enable modem support if the retail Network Adaptor (which includes a built-in V.90 modem) is used. The kit supports display on RGB monitors (with sync-on-green) using a VGA cable provided with the Linux Kit, or television sets with the normal cable included with the PlayStation 2 unit.

The PS2 Linux distribution is based on Kondara MNU/Linux, a Japanese distribution itself based on Red Hat Linux. PS2 Linux is similar to Red Hat Linux 6, and has most of the features one might expect in a Red Hat Linux 6 system. The stock kernel is Linux 2.2.1, but it can be upgraded to a newer version such as 2.2.21, 2.2.26 or 2.4.17.

Open-source applications

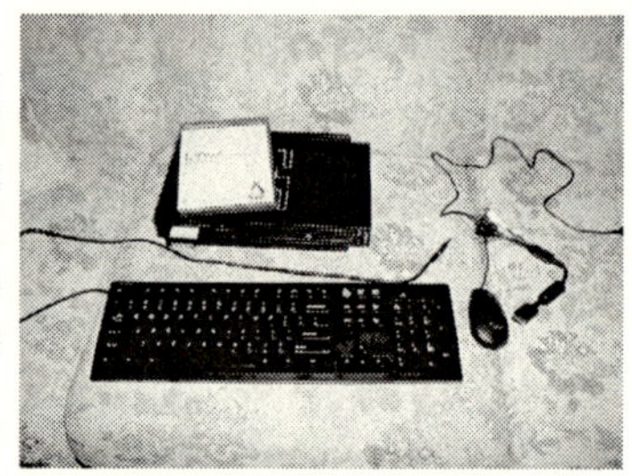
Contents of the Linux kit.

The Linux kit's primary purpose is amateur software development, but it can be used as one would use any other computer, although the small amount of memory in the PS2 (32MB) limits its applications. Noted open source software that compiles on the kit includes Mozilla Suite, XChat, and Pidgin. Lightweight applications better suited to the PS2's 32MB of RAM include xv, Dillo, Ted, and AbiWord. The default window manager is Window Maker, but it is possible to install and use Fluxbox and FVWM. The USB ports of the console can be connected to external devices, such as printers, cameras, flash drives, and CD drives.

With PS2 Linux, a user can program his/her own games that will work under PS2 Linux, but not on an unmodified PlayStation 2. Free open source code for games are available for download from PS2 Linux support sites. There is little difference between PS2 Linux and the Linux software used on the more expensive system ("Tool") used by professional licenced PlayStation game programmers. Some amateur-created games are submitted to a competition such as the Independent Games Festival's annual competition. It is possible for an amateur to sell games or software that he/she develops using PS2 Linux, with certain restrictions detailed in the End User License Agreement. The amateur cannot make and sell game CDs and DVDs, but can sell the game through an online download.

Distribution

Playstation 2 Linux DVD cover.

As of 2003, this kit is no longer officially sold in the USA due to the entire allocation of NTSC kits being sold out, but it is available through import or through an auction site, such as eBay. Some incorrectly speculate it was used as an attempt to help classify the PS2 as a computer to achieve tax exempt status from certain EU taxes that apply to game consoles and not computers (It was the Yabasic included with EU units that was intended to do that).Wikipedia:Citation needed Despite this, Sony lost the case in June 2006. The kit was released in the spirit of the earlier, Net Yaroze PlayStation and Sony continued their support of hobbyist programmers with the support of Linux on the PlayStation 3.

Model compatibility

The original version of the PS2 Linux kit only worked on the Japanese SCPH-10000, SCPH-15000 and SCPH-18000 Playstation 2 models. It came with a PCMCIA interface card which had a 10/100 Ethernet port and an external IDE hard drive enclosure (as there is no room inside the unit). This kit cannot be used with any later model PS2 (which includes all non-Japanese models) because these models removed the PCMCIA port.

Later versions of the PS2 Linux kit use an interface very similar to the HDD interface/ethernet sold later for network play (the later released Network adaptor was also usable with the kit, including the built-in 56k modem.) This kit locates the hard drive internal to the PS2, in the MultiBay. With this kit, only the SCPH-30000 model of PlayStation

2 is officially supported. The kit does though work equally well with models newer than SCPH-30000 with the exception that the ethernet connection tended to freeze after a short period of use. Thus the newer SCPH-50000 PlayStation 2 model will only work correctly with PS2 Linux with an updated network adapter driver, which must be transferred to the PlayStation 2 HDD by using either an older model PlayStation 2 to transfer the driver or a Linux PC with an IDE port. Both methods involve swapping HDDs, and the latter method requires opening the PC's case. This is due to the inability to use USB Mass Storage devices with the relatively old kernel (version 2.2.1) shipped with the kit.

The slim SCPH-70000 PlayStation 2 model does not work with PS2 Linux at all, due to the lack of a hard drive interface, though a very few early models in this revision had solder pads of an IDE interface on the motherboard that could be used (but required modding of the console, thereby voiding its warranty.) Even so, it is possible to network boot from a PXE server

PS2 Linux install DVDs are region encoded, as are all other PS2 game discs. A European/PAL disc will be rejected by an NTSC PlayStation 2 game system, however this is only at boot time: if you have a legal mod that allows you to load a PAL disk, then the PS2 Linux boot loader supports both PAL and Linux (read the documentation to determine the button presses), so once you are past the "DVD not supported", you can boot Linux and then later start X Window in NTSC mode.

External links

- Sony's PlayStation2 Linux Community [3]

References

[1] http://playstation2-linux.com/forum/forum.php?forum_id=1251
[2] " PS2DEV.ORG: Playstation Programming (http://ps2dev.org/)". .

PCSX2

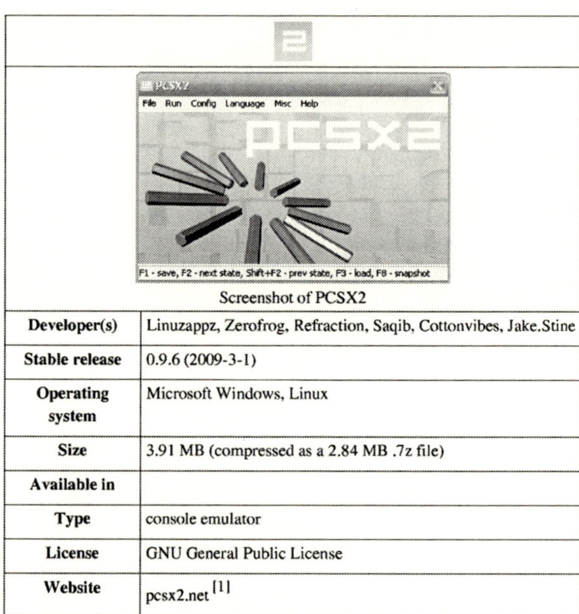

Screenshot of PCSX2

Developer(s)	Linuzappz, Zerofrog, Refraction, Saqib, Cottonvibes, Jake.Stine
Stable release	0.9.6 (2009-3-1)
Operating system	Microsoft Windows, Linux
Size	3.91 MB (compressed as a 2.84 MB .7z file)
Available in	
Type	console emulator
License	GNU General Public License
Website	pcsx2.net [1]

PCSX2 is a → PlayStation 2 (PS2) emulator for the Microsoft Windows and Linux operating systems. With the most recent versions, many PS2 games are playable (although speed limitations have made play-to-completion tests for many games impractical), and several games are claimed to have full functionality.[2] The main bottleneck in PS2 emulation is emulating the multi-core PS2 on PC x86 architecture. Although each CPU can be emulated well, getting the synchronization and timing between them to be accurate is very difficult.

PCSX2, like its predecessor project PCSX, is based on a plugin architecture, separating several functions from the core emulator. These are the graphics, controls, CD/DVD drive, USB, sound and FireWire (i. Link) ports. Different plugins may produce different results in both compatibility and performance. Additionally, PCSX2 requires a copy of the PS2 BIOS to operate, which is not offered for download by the developer, due to the copyright concerns and legal issues associated with it. A free clone of the official BIOS is in development.

Performance

Speed prior to version 0.9 was between 2 and 15 frames per second depending on the game, PC hardware, and plugin configuration—with the PS2 BIOS itself being one of the fastest pieces of software to emulate. Since the release of version 0.9 in April 2006, performance has greatly increased. Most 2D games and menus can reach 60-120 frame/s when specific plugins are used, and with the latest version, in-game 3D performance on a relatively new desktop computer can reach speeds greater than the native PS2 frame rate of 60 frame/s (NTSC) and 50 frame/s (PAL). In version 0.9, PCSX2 also began supporting dual core CPUs, resulting in a very significant increase in frame rate on systems with dual core processors. As of 0.9.1 in summer 2006, dual-core CPUs (Athlon 64 X2 and Core 2 Duo running at approximately 3.0 GHz) can run popular 3D games, such as Final Fantasy X, at well over 60 frame/s (when combined with an adequate video card such as a Radeon X1000 series or GeForce 6+ series, or midrange

Radeon HD2600xt or Geforce 8600 card). The developers and beta testers claim that Final Fantasy X is fully playable to completion. The PCSX2 team continues to remain actively involved in the development of PCSX2.

Misconceptions regarding speed

See also: megahertz myth

A common misconception regarding PCSX2 is that a processor with a relatively high clock speed, such as a Pentium 4 or Athlon XP (the former with clock speeds as high as 3.8 GHz), should easily be able to emulate PS2 games at full speed. The PS2 has several processing units including a MIPS R5900 chip, MIPS R3000A chip, two custom vector units, and graphics chip (Graphics Synthesizer). With the exception of the MIPS R5900 (clocked at 294.912 MHz) and the MIPS R3000A (clocked at 36.864 MHz, selectable to 33.8688 MHz for PlayStation (1) emulation), all other chips run at the bus speed of 147 MHz. There are several reasons which make emulation of the PS2 on a PC extremely difficult. Architectural differences between an x86-based PC and the PS2 are substantial; emulating multiple chips running in parallel on a single-core processor is quite complex. Taking advantage of dual core processors on PCs is even harder due to the tight synchronization between the PS2 chips. The development team provides a blog, explaining technical details of emulating the PS2.[3]

Plugin Development

Several plugins are currently being further developed, for performance and compatibility enhancements.

GSdx[4] is the highest performing graphics plugin and utilizes DirectX technology. Development is tied to the Google Code project, so new builds will follow in step with PCSX2's project SVN numbers. Hardware mode is the fastest and enables a texture resolution hack that can upscale to virtually any amount (thus creating a smoother lines and less artifacts), but this mode is not always the most accurate, and at times switching to the native resolution or the software mode will create a more complete image. Software mode is necessarily slower, but it has the option of choosing the number of processor threads, which is optimally the maximum number of cores your processor contains.

ZZogl[5] (based on Zero GS) is the most recently developed OpenGL-based graphics plugin, which is commonly used by Linux users considering DirectX is for Windows only. It is often noticeably slower than GSdx, particularly because it isn't efficiently offloading enough tasks to the video card.

Netplay

Recently, Gigaherz (one of the PCSX2's developers), has been working on allowing the emulator to connect to the PS2 internet service, enabling certain games to be playable over the internet with other players, even if they are playing on a PS2 console. The two games which the PCSX2 team have confirmed to be working with the netplay function on version 0.9.4 are XIII and Monster Hunter, but there may be more.

System requirements

Minimum

- Windows XP/Vista or Linux 32bit/64bit
- CPU that supports SSE2 (Pentium 4 and up, Athlon64 and up)
- GPU that supports Pixel Shader 2.0
- 512MB RAM

Recommended

- Windows XP/Vista or Linux 32bit/64bit
- CPU: Core 2 Duo 3.2GHz
- GPU: GeForce 8600 GT or better
- 1GB RAM (2GB if using Windows Vista)

External links

- Official website [6]
- PCSX2 project page [7] on Google Code
- The Official PCSX2 Archive [8], a collection of PCSX2 releases and Plugins
- irc://irc.efnet.org/pcsx2, unofficial PCSX2 support channel on IRC.

References

[1] http://www.pcsx2.net/
[2] " Compatibility (http://www.pcsx2.net/compat.php)". PCSX2. .
[3] " Blog (http://www.pcsx2.net/blog.php)". PCSX2. .
[4] http://forums.pcsx2.net/thread-3031.html
[5] http://forums.pcsx2.net/thread-4108.html
[6] http://www.pcsx2.net
[7] http://code.google.com/p/pcsx2/
[8] http://www.generalemu.net/pcsx2archive/

List of PlayStation 2 games

This is a list of about 1800 games for the → PlayStation 2 video game system, both released and unreleased

A

Title and source	Year	Developer(s)	Publisher(s)	Regions released
Abidoo and the Energy Thieves[1]	2004	Coktel Vision	VU Games	EU
Ace Combat 04: Shattered Skies	2001	Namco	Namco	
Ace Combat 5: The Unsung War	2004	Namco	Namco	
Ace Combat Zero: The Belkan War	2006	Namco	Namco	
Ace Lightning[1]	2002	BBC Multimedia	BBC Multimedia	
Action Girlz Racing	2005	Data Design Interactive	Metro 3D	EU
Activision Anthology	2002	Activision	Contraband Entertainment	NA EU
ADK Tamashii[1]	2008	SNK Playmore	SNK Playmore	JP
The Adventures of Cookie & Cream	2000	Agetec	From Software	JP NA EU
The Adventures of Jimmy Neutron Boy Genius: Attack of the Twonkies	2004	THQ	THQ	NA EU AUS
Æon Flux	2005	Majesco	Terminal Reality	NA
Aero Elite: Combat Academy	2003	Sega	Sega-AM2	

AFL Live 2003	2002	IR Gurus	Acclaim	AUS
AFL Live 2004	2003	IR Gurus	Acclaim	AUS
AFL Live Premiership Edition	2004	IR Gurus	Acclaim	AUS
AFL Premiership 2005	2005	IR Gurus	SCE	AUS
AFL Premiership 2006	2006	IR Gurus	SCE	AUS
AFL Premiership 2007	2007	IR Gurus	SCE	AUS
Agassi Tennis Generation	2003	Dreamcatcher Interactive	Aqua Pacific	
Age of Empires II: The Age of Kings	2001	Konami	Microsoft	
Agent Hugo[1]	2005	ITE Media	ITE Media	EU
Agent Hugo: Hula Holiday[1]	2008	NDS Software	NDS Software	EU
Agent Hugo - Lemoon Twist[1]	2007	NDS Software	NDS Software	EU
Agent Hugo: Roborumble[1]	2006	ITE Media	ITE Media	EU
Aggressive Inline	2002	Z-Axis	Acclaim	NA EU AUS
Air	2002	Key	NEC Interchannel	JP
Air Raid 3[1]	2004	Phoenix Games	Phoenix Games	EU AUS
Air Ranger: Rescue Helicopter	2001	ASK Digital	Midas Interactive Entertainment	JP EU
Air Ranger: Rescue Helicopter 2	2001	ASK Digital	Midas Interactive Entertainment	
AirBlade	2001	Criterion Games	Namco	NA EU AUS
Airborne Troops: Countdown to D-Day[1]	2005	Playlogic International	Widescreen Games	
Airforce Delta Strike	2004	Konami	Konami	JP NA EU
The Aka-Champion[1]	2005	D3 Publisher	D3 Publisher	JP
Akira Psychoball	2002	Bandai	Bandai	JP
Akudaikan[1]		Global A	Global A	JP
Alarm for Cobra 11: Hot Pursuit	2005	Midas Interactive	Midas Interactive	EU
Alex Ferguson's Player Manager 2001	2001	Anco	3DO	EU
Alex Ferguson's Player Manager 2002	2002	Anco	3DO	EU
Alias	2004	Acclaim	Acclaim	
Alien Hominid[1]	2004	O3 Entertainment	The Behemoth	NA EU
Aliens versus Predator: Extinction	2003	Zono Inc.	Electronic Arts	NA EU
All Star Pro-Wrestling	2000	Square	Square	JP
All Star Pro-Wrestling II	2001	Square Co., Ltd.	Square Co., Ltd.	JP
All-Star Baseball 2002	2001	Acclaim Entertainment	Acclaim Entertainment	NA EU
All-Star Baseball 2003	2002	Acclaim Entertainment	Acclaim Entertainment	NA EU
All-Star Baseball 2004	2003	Acclaim Entertainment	Acclaim Entertainment	NA EU
All-Star Baseball 2005	2004	Acclaim Entertainment	Acclaim Entertainment	NA EU
The All Star Kakutou	2005	Digital Zero	D3 Publisher	JP
Alfa Romeo Racing Italiano	2006	Milestone S.r.l.	Valcon Games	NA

Alpine Racer 3	2002	Namco	Namco	JP
Alpine Skiing 2005	2005	49Games	Midas Interactive Entertainment	EU
Alter Echo	2003	Outrage Entertainment	THQ	
Alone in the Dark: The New Nightmare	2001	Darkworks, Spiral House	Infogrames	NA
America's 10 Most Wanted	2004	Play It	Play It	EU
American Chopper	2004	Creat Studios	Activision	
American Chopper: Full Throttle	2005	Creat Studios	Activision	
American Idol	2003	Hothouse Creations	Codemasters	NA EU
AMF Xtreme Bowling 2006	2006	Atomic Planet	Mud Duck Productions	
Amplitude	2003	Harmonix	SCEI	
AND 1 Streetball	2006	Black Ops	Ubisoft	
Angel Present: A Marl Kingdom Story	2000	Nippon Ichi	Nippon Ichi	JP
Another Century's Episode	2005	From Software	Banpresto	JP
Another Century's Episode 2	2006	From Software	Banpresto	JP
Another Century's Episode 3	2007	From Software	Banpresto	JP
The Ant Bully	2006	A2M	Midway Games	
Antz Extreme Racing	2002	Supersonic Software	Empire Interactive, Light and Shadow Productions	
Aoi Umi no Tristia: Nanoka Franka Hatsumei Koubouki	2005	NIS	Kogado Software Products	JP
Ape Escape: Million Monkeys	2006	SCEI	SCEI	JP
Ape Escape: Pumped & Primed	2004	SCEI	SCEI, Ubisoft	
Ape Escape 2	2003	SCEI	SCEI, Ubisoft	
Ape Escape 3	2006	SCEI	SCEI	
Aqua Aqua	2000	Zed Two	SCi, 3DO	
Arcana Heart	2007(JP) 2008(NA)	Examu	AQ Interactive (JP) Atrus (NA)	JP NA
Arc the Lad: End of Darkness	2005	Cattle Call	SCEI, Namco	
Arc the Lad: Twilight of the Spirits	2003	Cattle Call	SCEI	JP NA EU
Arctic Thunder	2001	Midway	Midway	
Area 51	2005	Midway Studios Austin	Midway	
Arena Football	2006	Electronic Arts	EA Sports	
Arena Football: Road to Glory	2007	EA Tiburon, Budcat Creations	EA Sports	
Armored Core 2	2000	From Software	Agetec	
Armored Core 2: Another Age	2001	From Software	Agetec	
Armored Core 3	2002	From Software	Agetec	
Armored Core: Formula Front	2005	From Software	Agetec	JP
Armored Core: Nine Breaker	2005	From Software	Agetec	
Armored Core: Last Raven	2006	From Software	Agetec	

Army Men: Air Attack 2	2001	3DO	3DO	
Army Men: RTS	2002	Pandemic Studios	3DO	
Army Men: Sarge's Heroes 2	2001	3DO	3DO	
Ar tonelico: Melody of Elemia	2006	Gust	Banpresto, NIS America	JP NA EU
Ar tonelico II: Melody of Metafalica	2007	Gust	Banpresto, NIS America	JP NA EU
Asterix & Obelix Kick Buttix	2004	Etranges Libellules	Atari	
Asterix & Obelix XXL 2	2006	Etranges Libellules	Atari	EU
Asterix at the Olympic Games	2007	Etranges Libellules	Atari	EU
Astro Boy	2004	Sonic Team	Sega	
Atari Anthology	2004	Digital Eclipse Software	Atari	
Atelier Iris: Eternal Mana	2005	Gust	Gust, NIS, Koei	JP NA EU
Atelier Iris 2: The Azoth of Destiny	2006	Gust	Gust, NIS, Koei	JP NA EU
Atelier Iris: Grand Fantasm	2007	Gust	Gust, NIS, Koei	JP NA EU
Athens 2004	2004	Eurocom	SCEI	
Atlantis III: The New World	2001	Cryo Interactive Entertainment	Cryo Interactive Entertainment	
A Train 6	2000	Artdink	Midas Interactive	JP
ATV Offroad Fury	2001	Rainbow Studios	SCEI	
ATV Offroad Fury 2	2002	Rainbow Studios	SCEI	
ATV Offroad Fury 3	2004	Climax Group	SCEI	
ATV Offroad Fury 4	2006	Climax Group	SCEI	
ATV Quad Power Racing 2	2003	Climax Group	Acclaim Entertainment	
Auto Modellista	2003	Capcom Production Studio 1	Capcom	
Avatar: The Last Airbender	2006	THQ	THQ	

B

Title and source	Year	Developer(s)	Publisher(s)	Regions released
Babe	2006	Aqua Pacific	Blast! Entertainment	EU
Backyard Baseball	2004	Humongous Entertainment	Atari	NA
Backyard Baseball 2007	2006	Humongous Entertainment	Humongous Entertainment	NA
Backyard Baseball 2009	2008	Humongous Entertainment	Humongous Entertainment	NA
Backyard Basketball	2003	Humongous Entertainment	Atari	NA
Backyard Football	2005	Humongous Entertainment	Atari	NA
Backyard Football '08	2007	Humongous Entertainment	Atari	NA
Backyard Football '09	2008	Humongous Entertainment	Atari	NA
Backyard Wrestling: Don't Try This at Home	2003	Paradox Development	Eidos	NA EU
Backyard Wrestling 2: There Goes the Neighborhood	2004	Paradox Development	Eidos	NA EU

Bad Boys: Miami Takedown	2004	Blitz Games	Crave Entertainment	NA EU
Bakushou! Jinsei Kaimichi	2004	Taito	Taito	JP
Baldur's Gate: Dark Alliance	2001	Snowblind Studios	Interplay	JP NA EU
Baldur's Gate: Dark Alliance II	2004	Black Isle Studios	Interplay	NA
Bakufu Slash!!Kizna Arashi	2004	SCE Japan Studio	Sony Computer Entertainment	JP
Barbarian	2002	Saffire	Titus	JP NA EU
Barbie Horse Adventures: Wild Horse Rescue	2003	Blitz Games	Vivendi	NA
The Bard's Tale	2004	InXile Entertainment	Vivendi	NA
Barnyard	2006	Blue Tongue	THQ	NA EU
Bass Strike	2001	Pai inc.	THQ	NA EU
Batman Begins	2005	Eurocom	EA Games	NA EU
Batman: Rise of Sin Tzu	2003	Ubisoft Montreal	Ubisoft	NA EU
Batman Vengeance	2001	Ubisoft Montreal	Ubisoft	NA
Battle Engine Aquila	2003	Lost toys games	Atari	NA EU
Battlefield 2: Modern Combat	2005	DICE	EA Games	JP NA EU
Battle Stadium D.O.N	2006	8ing/Raizing	Namco Bandai	JP
Battlestar Galactica	2003	Warthog	Sierra	
BCV: Battle Construction Vehicles	2000	Artdink	Midas Interactive Entertainment	JP EU
Beat Down: Fists of Vengeance	2005	Cavia	Capcom	
Beatmania	2006	Konami	Konami	NA
Beatmania IIDX 3rd Style	2000	Konami Computer Entertainment Japan	Konami	JP
Beatmania IIDX 4th Style: New Songs Collection	2001	Konami Computer Entertainment Japan	Konami	JP
Beatmania IIDX 5th Style: New Songs Collection	2001	Konami Computer Entertainment Japan	Konami	JP
Beatmania IIDX 6th Style: New Songs Collection	2002	Konami Computer Entertainment Japan	Konami	JP
Beatmania IIDX 7th Style	2004	Konami Computer Entertainment Studios	Konami	JP
Beatmania IIDX 8th Style	2004	Konami Computer Entertainment Studios	Konami	JP
Beatmania IIDX 9th Style	2005	Konami Computer Entertainment Studios	Konami	JP
Beatmania IIDX 10th Style	2005	Konami	Konami	JP
Beatmania IIDX 11: IIDX RED	2006	Konami Digital Entertainment	Konami Digital Entertainment	JP
Beatmania IIDX 12: HAPPY SKY	2006	Konami Digital Entertainment	Konami Digital Entertainment	JP
Beatmania IIDX 13: DisorteD	2007	Konami Digital Entertainment	Konami Digital Entertainment	JP
Beatmania IIDX 14: GOLD	2008	Konami Digital Entertainment	Konami Digital Entertainment	JP
Beatmania IIDX 15: DJ TROOPERS	2008	Konami Digital Entertainment	Konami Digital Entertainment	JP
Berserk	2004	Sammy	Sammy	JP
Beyond Good and Evil	2003	Ubisoft Montpellier	Ubisoft	NA EU

The Bible Game	2005	Crave Entertainment	Crave Entertainment	NA EU
Big Mutha Truckers	2002	Eutechnyx	Empire Interactive/THQ	NA EU
Big Mutha Truckers 2	2005	Empire Interactive	THQ	NA EU
Bionicle Heroes	2006	TT Games, Amaze Interactive	Eidos Interactive	JP NA EU
Bionicle: The Game	2003	Argonaut Games	Lego Media/Electronic Arts	NA
Black	2006	Criterion Games	Electronic Arts	JP NA EU
Black & Bruised	2003	Digital Fiction	Majesco	NA
Black Market Bowling	2005	Black Market Games	Midas Interactive Entertainment	EU
Blade II	2002	Mucky Foot	Activision	
Bleach: Blade Battlers	2006	Racjin	SCEI	JP
Bleach: Blade Battlers 2	2007	Racjin	SCEI	JP
Bleach: Erabareshi Tamashii	2005	SCEI	SCEI	JP
Bleach: Hanatareshi Yabou	2006		SCEI	JP
Blitz: The League	2005	Midway Games	Midway Games	
Blood Omen 2	2002	Crystal Dynamics	Eidos Interactive	
Blood+: One Night Kiss	2006	Grasshopper Manufacture	Namco Bandai	JP
Blood: The Last Vampire - First Volume	2000	Sugar & Rockets	SCEI	JP
Blood: The Last Vampire - Last Volume	2000	Sugar & Rockets	SCEI	JP
Blood Will Tell	2004	Sega	WOW Entertainment	JP NA EU
BloodRayne	2002	Terminal Reality	Majesco	
BloodRayne 2	2004	Terminal Reality	Majesco	
Bloody Roar 3	2001	Eighting	Activision	NA
Bloody Roar 4	2003	Eighting	Activision	NA EU
BlowOut	2003	KaosKontrol	Majesco	
BMX XXX	2002	Z-Axis	Acclaim Entertainment	NA
Bobobo-bo Bo-bobo Atsumare! Taikan Bo-bobo	2004	Hudson	Hudson	JP
Bobobo-bo Bo-bobo Hajike Matsuri	2003	Hudson	Hudson	JP
Bode Miller Alpine Skiing	2006	49Games	Valcon Games	
Boku no Natsu Yasumi 2	2004	Millennium Kitchen	SCEI	JP
Bombastic	2003	Shift	Capcom	
Bomberman Hardball	2005	Hudson	Ubisoft	JP EU
Bomberman Jetters	2003	Hudson	Hudson	JP
Bomberman Land 2	2003	Racjin	Hudson	JP
Bomberman Kart	2001	Racjin, Hudson	Hudson	JP
Bouken Jidai Katsugeki: Goemon	2000	Konami	Konami	JP
The Bouncer	2000	Dream Factory	Squaresoft	JP NA EU
Boxing Champions	2002	Tamsoft	Midas Interactive Entertainment	JP
Bratz: Forever Diamondz	2006	Blitz Games	THQ	
Bratz: Rock Angelz	2005	Blitz Games	THQ	

Brave: The Search For Spirit Dancer	2007	Vis Entertainment	Evolved Games	
Bravo Music: Chou-Meikyokuban	2002	Desert Productions, SCEI	SCEI	JP
Bravo Music: Christmas Edition	2001	Desert Productions, SCEI	SCEI	JP
Breath of Fire: Dragon Quarter	2003	Capcom Production Studio 3	Capcom	
Brian Lara International Cricket	2005	Swordfish Studios	Codemasters	EU
Britney's Dance Beat	2002	Metro Graphics	THQ	NA
Broken Sword: The Sleeping Dragon	2003	Revolution Software	THQ	JP EU
Brothers in Arms: Earned in Blood	2005	Gearbox Software	Ubisoft	NA
Brothers In Arms: Road to Hill 30	2005	Gearbox Software	Ubisoft	NA
Buffy the Vampire Slayer: Chaos Bleeds	2003	Eurocom	Vivendi Games	NA EU
Bujingai: The Forsaken City	2003	Red Entertainment	Taito	JP NA EU
Bully (*Canis Canem Edit* in Europe/Australia)	2006	Rockstar Vancouver	Rockstar Games	NA EU
Burnout	2001	Criterion Games	Acclaim	NA EU
Burnout 2: Point of Impact	2002	Criterion Games	Acclaim	NA EU
Burnout 3: Takedown	2004	Criterion Games	EA Games	NA EU
Burnout Revenge	2005	Criterion Games	EA Games	NA EU
Burnout Dominator	2007	Criterion Games	EA Games	NA EU
Bust-A-Bloc	2002	Tamsoft	Midas Interactive Entertainment	JP
Butt Ugly Martians: Zoom or Doom!	2003	Runecraft	Crave Entertainment, Vivendi Games	
Buzz! The Big Quiz	2006	Relentless Software	SCEE	EU
Buzz! The Hollywood Quiz	2007	Relentless Software	SCEE	EU
Buzz! The Mega Quiz	2008	Relentless Software	SCEE	EU
Buzz! The Music Quiz	2005	Relentless Software	SCEE	EU
Buzz! The Sports Quiz	2006	Kuju Entertainment	SCEE	EU

C

Title and source	Year	Developer(s)	Publisher(s)	Regions released
Cabela's Alaskan Adventure	2006	Activision	Activision	NA
Cabela's Big Game Hunter (2002)	2002	Sand Grain Studios	Activision	NA
Cabela's Big Game Hunter (2007)	2007	Sand Grain Studios	Activision	NA
Cabela's Big Game Hunter 2005 Adventures	2004	Sand Grain Studios	Activision	NA
Cabela's Dangerous Hunts	2003	Sand Grain Studios	Activision	NA
Cabela's Dangerous Hunts 2	2005	Sand Grain Studios	Activision	NA
Cabela's Dangerous Hunts 2009	2008	Activision	Activision	NA
Cabela's Legendary Adventures	2008	Activision	Activision	NA
Cabela's Monster Bass	2007	Activision	Activision	NA
Cabela's Outdoor Adventures	2005	Sand Grain Studios	Activision	NA

Cabela's Trophy Bucks	2007	Sand Grain Studios	Activision	NA
Call of Duty: Finest Hour	2004	Spark Unlimited	Activision	
Call of Duty 2: Big Red One	2005	Treyarch, Gray Matter Studios	Activision	
Call of Duty 3	2006	Treyarch, Pi Studios	Activision	
Call of Duty: World at War: Final Fronts	2008	Rebellion	Activision	
Capcom Classics Collection	2005	Digital Eclipse Software	Capcom	
Capcom Classics Collection Vol. 2	2006	Digital Eclipse Software	Capcom	
Capcom Fighting Evolution	2004	Capcom	Capcom	
Captain Tsubasa	2006	Bandai	Bandai	JP
Carmen Sandiego: The Secret of the Stolen Drums	2004	A2M	BAM! Entertainment	
Cars	2006	Rainbow Studios	THQ	
Cartoon Network Racing	2006	Eutechnyx	Game Factory	
Castle Shikigami 2 Shikigami no Shiro 2[JP]	2004	Alfa System	XS Games	
Castlevania: Curse of Darkness	2005	Konami TYO	Konami	
Castlevania: Lament of Innocence	2003	Konami TYO	Konami	
Catwoman	2004	Electronic Arts UK, Argonaut Games	Electronic Arts	
Cel Damage Overdrive	2002	Pseudo Interactive	Electronic Arts	
Celebrity Deathmatch	2003	Big Ape Productions	Gotham Games	
Centre Court Hard Hitter				
Champions of Norrath: Realms of Everquest	2004	Snowblind Studios	Sony Online Entertainment	NA EU
Champions: Return to Arms	2005	Snowblind Studios	Sony Online Entertainment	
Championship Manager 2006	2006	Beautiful Game Studios	Eidos Interactive	EU
Championship Manager 2007	2007	Beautiful Game Studios	Eidos Interactive	EU
Championship Manager 5	2005	Beautiful Game Studios	Eidos Interactive	EU
Charlie and the Chocolate Factory	2005	High Voltage Software	Global Star Software	
Charlie's Angels	2003	Neko Entertainment	Ubisoft	
Chaos Legion	2003	Capcom Production Studio 6	Capcom	
The Chronicles of Narnia: The Lion, the Witch and the Wardrobe	2005	Traveller's Tales	Buena Vista Games	JP NA EU AUS
Chessmaster	2003	Ubisoft	Ubisoft	
ChoroQ		Takara		
Chulip	2002	Punchline	Victor Interactive, Natsume	JP NA EU
C.I.D 925: An Ordinary Life	2009	Oxygen Games, Twelve Games	Oxygen Games	NA EU
Circus Maximus: Chariot Wars	2002	Kodiak Interactive	THQ	NA EU
City Crisis	2001	Syscom Entertainment	Take-Two Interactive	
Clannad	2006	Key	Interchannel	JP
Clever Kids-Dino Land				

Clever Kids-Pony World				
Clock Tower 3	2002	Sunsoft	Capcom Production Studio 3	
Club Football AC Milan 2005				
Club Football Barcelona 2005				
Club Football Birmingham City 2005				
CMT Presents: Karaoke Revolution Country	2006	Harmonix	Konami	
Cocoto Platform Jumper	2004	Neko Entertainment	Big Ben Interactive	
Code Age Commanders	2005	Square Enix	Square Enix	
Code Lyoko: Quest for Infinity	2008	Neko Entertainment	The Game Factory	
Code of the Samurai				
Cold Fear	2005	Darkworks	Ubisoft	
Cold Winter	2005	Swordfish Studios	Vivendi	
Colin McRae Rally 3	2003	Codemasters	Codemasters	NA EU
Colin McRae Rally 04	2003	Codemasters	Codemasters	
Colin McRae Rally 2005	2004	Codemasters	Codemasters	
College Hoops 2K6	2005	Visual Concepts	2K Sports	
College Hoops 2K7	2006	Visual Concepts	2K Sports	
College Hoops 2K8	2007	Visual Concepts	2K Sports	
Coloball 2002	2002	Vanpool	Enterbrain	JP
Colosseum: Road to Freedom	2005	Ertain	KOEI	JP NA EU AUS
Combat Elite: WWII Paratroopers	2005	BattleBorne Entertainment	SouthPeak Interactive	
Commandos: Strike Force	2006	Pyro Studios	Eidos Interactive	
Commandos 2: Men of Courage	2002	Pyro Studios	Eidos Interactive	NA EU
Conflict: Desert Storm	2002	Pivotal Games	SCi	JP NA EU AUS
Conflict: Desert Storm II: Back to Baghdad	2003	Pivotal Games	SCi	JP NA EU AUS
Conflict: Vietnam	2004	Pivotal Games	SCi, Global Star Software	JP NA EU AUS
Conflict Zone	2002	MASA Group	Ubisoft	
Constantine	2005	Bits Studios	SCi, THQ	
Contra: Shattered Soldier	2002	Konami	Konami	
Convenience Store Manager 3				
Cookie and Cream				
Coraline	2009	Papaya Studio	D3 Publisher	NA
Cowboy Bebop	2005	Banpresto	Bandai	
Crash Bandicoot: The Wrath of Cortex	2001	Traveller's Tales	Universal Interactive Studios	JP NA EU AUS
Crash Nitro Kart	2003	Vicarious Visions	Vivendi	JP NA EU AUS
Crash Tag Team Racing	2005	Radical Entertainment	Sierra Entertainment	JP NA EU AUS
Crash Twinsanity	2004	Traveller's Tales	Vivendi	JP NA EU AUS
Crash of the Titans	2007	Radical Entertainment	Sierra Entertainment	NA EU AUS

Crash: Mind Over Mutant	2008	Radical Entertainment	Activision	NA EU AUS
Crash 'n Burn				
Crazy Frog Racer				EU
Crazy Frog Racer 2				EU
Crazy Taxi	1999	Acclaim Studios Cheltenham	Sega	
Crescent Suzuki Racing: Superbikes and Supersides				
Cricket		EA Sports		
Crime Life: Gang Wars				
Crimson Sea 2		KOEI		
Crimson Tears		Spike		
Cross Channel: To All People				
Crouching Tiger, Hidden Dragon		Genki		
Crusty Demons				
Crystal Zone				
CSI: 3 Dimensions of Murder	2007	Telltale Games	Ubisoft	NA
Cue Academy				
Culdcept	2003	OmiyaSoft	NEC Interchannel	NA
Curious George				
Cyber Mahjong				
Cyclone Circus				
Cy Girls		Konami		

D

Title and source	Year	Developer(s)	Publisher(s)	Regions released
D1 Grand Prix				
Daemon Summoner				
Dance Dance Revolution Disney Channel Edition	2008	Keen Games	Konami Digital Entertainment	NA
Dance Dance Revolution Extreme	2003	Konami Computer Entertainment Japan	Konami	JP
Dance Dance Revolution Extreme	2004	Konami Digital Entertainment	Konami Digital Entertainment	NA
Dance Dance Revolution Extreme 2	2005	Konami Digital Entertainment	Konami Digital Entertainment	NA
Dance Dance Revolution Party Collection	2003	Konami Computer Entertainment Japan	Konami	JP
Dance Dance Revolution Strike	2006	Konami Digital Entertainment	Konami Digital Entertainment	JP
Dance Dance Revolution SuperNova	2007	Konami Digital Entertainment	Konami Digital Entertainment	JP
Dance Dance Revolution SuperNova	2006	Konami Digital Entertainment	Konami Digital Entertainment	NA
Dance Dance Revolution SuperNova 2	2008	Konami Digital Entertainment	Konami Digital Entertainment	JP
Dance Dance Revolution SuperNova 2	2007	Konami Digital Entertainment	Konami Digital Entertainment	NA
Dance Dance Revolution X	2008	Konami Digital Entertainment	Konami Digital Entertainment	JP NA

Dance Factory	2006			NA EU
Dance: UK	2003		Broadsword Interactive	
Dancing Stage Fever	2003	Konami Digital Entertainment GmbH	Konami Digital Entertainment GmbH	EU
Dancing Stage Fusion	2004	Konami Digital Entertainment GmbH	Konami Digital Entertainment GmbH	EU
Dancing Stage Max	2005	Konami Digital Entertainment GmbH	Konami Digital Entertainment GmbH	EU
Dancing Stage MegaMix	2003	Konami Digital Entertainment GmbH	Konami Digital Entertainment GmbH	EU
Dancing Stage SuperNova	2007	Konami Digital Entertainment GmbH	Konami Digital Entertainment GmbH	EU
Dancing Stage SuperNova 2	2008	Konami Digital Entertainment GmbH	Konami Digital Entertainment GmbH	EU
Dark Angel: Vampire Apocalypse			Metro 3D	
Dark Cloud	2000	Level-5	Sony Computer Entertainment	JP NA EU
Dark Cloud 2	2002	Level-5	Sony Computer Entertainment	JP NA EU
Dark Summit				
Darkwatch		High Moon Studios		
The Da Vinci Code				
DDR Festival Dance Dance Revolution	2004	Konami Computer Entertainment Japan	Konami	JP
DDRMAX Dance Dance Revolution	2002	Konami of America	Konami of America	NA
DDRMAX Dance Dance Revolution 6thMix	2002	Konami Computer Entertainment Tokyo	Konami	JP
DDRMAX2 Dance Dance Revolution	2003	Konami Digital Entertainment	Konami Digital Entertainment	NA
DDRMAX2 Dance Dance Revolution 7thMix	2003	Konami Computer Entertainment Japan	Konami	JP
Dead or Alive 2: Hardcore	2000	Team Ninja	Tecmo	JP NA EU
Dead Rush		Treyarch		
Dead To Rights			Namco	
Dead To Rights II			Namco	
Deadly Strike				
Death by Degrees			Namco	
Def Jam Fight for NY		Aki Corporation	Electronic Arts	
Def Jam Vendetta		Aki Corporation	Electronic Arts	
Defender			Midway Games	
Delta Force: Black Hawk Down				
Delta Force: Black Hawk Down: Team Sabre	2006			
Densha de go! Professional 2			Taito	
Densha de go! Final			Taito	
Destroy All Humans!	2005	Pandemic Studios	THQ	JP NA EU
Destroy All Humans! 2	2006	Pandemic Studios	THQ	NA EU
Destruction Derby Arenas		Studio 33/SCEE		

Deus Ex: The Conspiracy	2002	Ion Storm	Eidos Interactive	
Devil Kings			Capcom	
Devil May Cry		Capcom Production Studio 4	Capcom	
Devil May Cry 2		Capcom Production Studio 1	Capcom	
Devil May Cry 3: Dante's Awakening		Capcom Production Studio 1	Capcom	
Devil May Cry 3: Special Edition		Capcom Production Studio 1	Capcom	
Digimon Rumble Arena 2			Bandai	
Digimon World 4			Bandai	
Dirge of Cerberus: Final Fantasy VII	2006	Square Enix	Square Enix	JP NA EU
Dirttrack Devils				
Disaster Report		Irem		
Disaster Report 2		Irem		
Disgaea: Hour of Darkness				
Disgaea 2: Cursed Memories				
Disney's Extreme Skate Adventure				
Disney Golf		T&E Soft		
DNA Dark Native Apostle		Virgin interactive	Hudson Soft	
DoDonPachi Dai Ou Jou		Cave		
Dog's Life		Frontier Developments/SCEE		
Dora the Explorer: Journey to the Purple Planet				
Downforce				
Downhill Domination				
Dragon Ball Z: Budokai				
Dragon Ball Z: Budokai 2				
Dragon Ball Z: Budokai 3				
Dragon Ball Z: Budokai Tenkaichi		Spike		
Dragon Ball Z: Budokai Tenkaichi 2		Spike		
Dragon Ball Z: Budokai Tenkaichi 3				
Dragon Ball Z: Infinite World				
Dragon Ball Z: Sagas		Avalanche Software		
Dragon Quest and Final Fantasy In Itadaki Street Special			Square Enix	JP
Dragon Quest V: Bride of the Sky	2004	Artepiazza	Square Enix	JP
Dragon Quest VIII: Journey of the Cursed King	2005	Level-5	Square Enix	JP NA EU
Dragon Quest Yangus	2006		Square Enix	JP
Dragon's Lair 3D: Return to the Lair		Ubisoft		
Drakan: The Ancients' Gates		Surreal Software		
Drakengard		Cavia	Square Enix	JP NA EU

Drakengard 2		Cavia	Square Enix	JP NA EU
DreamMix TV World Fighters	2003	Hudson Soft / Konami	Red Company	JP
DRIV3R		Reflections Interactive		
Driver: Parallel Lines				
Driven		BAM! Entertainment		
Driven to Destruction		Atari		
Driving Emotion Type-S		Escape		
Dr. Muto	2002	Midway Games	Midway Games	JP NA EU
Drome Racers				
Dropship: United Peace Force		SCE Studios Camden		
DrumMania	2000	Konami Computer Entertainment Japan	Konami	JP
DT Racer				
Dual Hearts		Cazworks Studio/Matrix Software		
Duel Masters		Microsoft Game Studios, High Voltage Software		
The Dukes of Hazzard: Return of the General Lee		Ratbag Games		
Dynasty Tactics	2002	Omega Force	Koei	
Dynasty Tactics 2	2003	Omega Force	Koei	
Dynasty Warriors 2	2000	Omega Force	Koei	
Dynasty Warriors 3	2001	Omega Force	Koei	
Dynasty Warriors 3: Xtreme Legends	2003	Omega Force	Koei	
Dynasty Warriors 4	2003	Omega Force	Koei	JP NA EU
Dynasty Warriors 4: Empires	2004	Omega Force	Koei	JP NA EU
Dynasty Warriors 4: Xtreme Legends	2003	Omega Force	Koei	JP NA EU
Dynasty Warriors 5	2005	Omega Force	Koei	
Dynasty Warriors 5: Empires	2006	Omega Force	Koei	
Dynasty Warriors 5: Xtreme Legends	2005	Omega Force	Koei	
Dynasty Warriors 6	2008	Omega Force	Koei	JP NA EU
Dynasty Warriors: Gundam	2007	Koei	Namco Bandai Games	JP
Dynasty Warriors: Gundam 2	2008	Koei	Namco Bandai Games	JP

E

Title and source	Year	Developer(s)	Publisher(s)	Regions released
Eagle Eye Golf				
Ecco the Dolphin: Defender of the Future				
Echo Night: Beyond				
Ed, Edd n Eddy: The Mis-Edventures				
eJay Clubworld		Crave Entertainment		

Endonesia	2001	Vanpool	Enix	JP
Energy Airforce		Taito		
Energy Airforce Aim Strike!		Taito		
Enter the Matrix	2003	Shiny Entertainment	Atari Konami	JP NA EU
Enthusia Professional Racing		Konami		
Ephemeral Fantasia		Konami		
Eragon		Sierra		
Escape from Monkey Island	2001	LucasArts	LucasArts	NA
Ethan Nelson's Adventure through Time and Space		E-Tech		
ESPN College Hoops	2003	Kush Games	Sega	
ESPN College Hoops 2K5	2004	Visual Concepts	Sega	
ESPN Major League Baseball	2004	Blue Shift	Sega	
ESPN National Hockey Night	2001	Konami	Konami	
ESPN NBA 2K5	2004	Visual Concepts	Sega	
ESPN NBA Basketball	2003	Visual Concepts	Sega	
ESPN NFL 2K5	2004	Visual Concepts	Sega	
ESPN NFL Football	2003	Visual Concepts	Sega	
ESPN NHL 2K5	2004	Kush Games	Sega	
ESPN NHL Hockey	2003	Kush Games	Sega	
ESPN X Games Skateboarding	2001	Konami	Konami	
Eternal Poison	2008	Flight-Plan	Atlus	JP NA
Eternal Quest		Tamsoft	Midas Interactive Entertainment	
Eternal Ring	2000	From Software	Agetec	JP NA EU
Euro 2004	2004	EA Canada	Electronic Arts	
European Tennis Pro	2003	Magical Company Ltd	Phoenix Games Ltd	EU
EOE: Eve of Extinction	2002	Yuke's	Eidos Interactive	JP NA EU
Everblue	2001	Arika	Capcom	JP EU
Everblue 2	2002	Arika	Capcom	JP NA EU
Evergrace	2000	From Software	Agetec	JP NA EU AUS
EverQuest Online Adventures	2003	SOE	SOE	NA EU
EverQuest Online Adventures: Frontiers	2003	SOE	SOE	
Everybody's Golf	2003	Clap Hanz	Sony Computer Entertainment	
Evil Dead: A Fistful of Boomstick	2003	Heavy Iron Studios	THQ	
Evil Dead: Regeneration	2005	Cranky Pants Games	THQ	
Evil Twin: Cyprien's Chronicles	2001	In Utero	Ubisoft	EU
Extreme Sprint 3010				EU
Extermination	2001	Deep Space	Sony Computer Entertainment	JP NA EU
EyeToy: AntiGrav	2004	Sony	Sony Computer Entertainment	
EyeToy: Chat	2005	Sony	Sony Computer Entertainment	NA EU

EyeToy: Groove	2003	Sony Computer Entertainment	Sony Computer Entertainment	JP NA EU AUS
EyeToy: Kinetic	2005	Sony Computer Entertainment	Sony Computer Entertainment	
EyeToy: Monkey Mania	2004	Sony Computer Entertainment	Sony Computer Entertainment	JP EU AUS
EyeToy: Operation Spy				
EyeToy: Play		Sony		
EyeToy: Play 2		Sony		
EyeToy: Play 3				

F

Title and source	Year	Developer(s)	Publisher(s)	Regions released
F1 Championship Season 2000	2000	Visual Science	EA Sports	EU
F1 2001	2001	Visual Science	EA Sports	EU
F1 2002	2002	Visual Science	EA Sports	EU
F1 Career Challenge	2003	Visual Science	EA Sports	EU
Fahrenheit	2005	Quantic Dream	Atari	NA EU
The Fairly OddParents: Shadow Showdown	2004	Blitz Games	THQ	
The Fairly OddParents: Breakin' Da Rules	2003	Blitz Games	THQ	
The Fairly OddParents: Gary Returns!				
Fallout: Brotherhood of Steel	2004	Interplay	Interplay	JP NA EU
Fame Academy: Dance Edition	2004	Nobilis	Monte Cristo	EU
Family Feud	2006	Atomic Planet	Global Star Software	
Family Game Night	2008	Electronic Arts	Electronic Arts	
Family Guy	2006	High Voltage Software	2K Games	
Fantastic Four	2005	7 Studios, Beenox Studios	Activision	
FantaVision	2000	SCEI	Sony Computer Entertainment	NA EU
The Fast and the Furious	2006	Eutechnyx	Namco Bandai Games	
Fatal Frame	2001	Tecmo	Tecmo	JP NA EU
Fatal Frame II: Crimson Butterfly	2003	Tecmo	Tecmo	JP NA EU
Fatal Frame III: The Tormented	2005	Tecmo	Tecmo	JP NA EU
Fatal Fury: Battle Archives Vol. 1	2006	SNK Playmore	SNK	NA
Fear & Respect		Edge of Reality	Midway Games	
Ferrari Challenge Trofeo Pirelli	2008	Eutechnyx	System 3	NA EU AUS
FIFA 2001	2000	EA Canada	EA Sports	NA EU
FIFA Football 2002	2001	EA Canada	EA Sports	NA EU
FIFA Football 2003	2002	EA Canada	EA Sports	NA EU
FIFA Football 2004	2003	EA Canada	EA Sports	NA EU
FIFA Football 2005	2004	EA Canada	EA Sports	NA EU
FIFA 06	2005	EA Canada	EA Sports	NA EU

FIFA 07	2006	EA Canada	EA Sports	JP NA EU
FIFA 08	2007	EA Canada	EA Sports	NA EU
FIFA 09	2008	EA Canada	EA Sports	NA EU AUS
FIFA 10	2009	EA Canada	EA Sports	NA EU AUS
FIFA Street	2005	EA Canada	EA Sports	NA EU
FIFA Street 2	2006	EA Canada	EA Sports	JP NA EU
FightBox	2004	Gamezlab	BBC Multimedia	EU
Fighting Angels				
Fighting Fury	2000	Tomy	Midas Interactive Entertainment	
Fight Club	2004	Genuine Games	Vivendi Universal Games	
Fight Night 2004	2004	EA Sports	EA Sports	
Fight Night Round 2	2005	EA Sports	EA Sports	JP NA EU
Fight Night: Round 3	2006	EA Sports	EA Sports	
Fighter Maker 2	2002	Enterbrain	Agetec	
Final Fantasy X	2001	Squaresoft	Squaresoft	JP NA EU
Final Fantasy XI	2004	Square Enix	Sony Computer Entertainment	JP NA
Final Fantasy XI: Rise of the Zilart	2004	Square Enix	Sony Computer Entertainment	JP NA
Final Fantasy XI: Chains of Promathia	2004	Square Enix	Sony Computer Entertainment	JP NA
Final Fantasy XI: Treasures of Aht Urhgan	2006	Square Enix	Sony Computer Entertainment	JP NA
Final Fantasy XI: Wings of the Goddess	2007	Square Enix	Sony Computer Entertainment	JP NA
Final Fantasy XII	2006	Square Enix	Square Enix	JP NA EU
Final Fantasy X-2	2003	Square Enix	Square Enix	JP NA EU
Final Fight: Streetwise	2006	Capcom Production Studio 8	Capcom	
Finding Nemo	2003	Traveller's Tales	THQ	JP NA EU AUS
Finny The Fish & The Seven Waters	2005	Sony Computer Entertainment	Natsume	
Firefighter F.D. 18	2004	Konami	Konami	
Fire Pro Wrestling Returns	2005	Spike		
First Kiss Stories	2003	HuneX	Broccoli	
Flatout	2004	Bugbear Entertainment	Empire Interactive, Konami	JP NA EU
Flatout 2	2006	Bugbear Entertainment	Empire Interactive, Konami	JP NA EU AUS
Flipnic: Ulitmate Pinball	2005	Sony Computer Entertainment	Capcom	
Flow: Urban Dance Uprising	2005	Artificial Mind and Movement	Ubisoft	
Flower, Sun, and Rain	2001	Grasshopper Manufacture	Victor Interactive, Marvelous Entertainment	JP
Flushed Away	2006	D3 Publisher	D3 Publisher	
Football Generation	2004	Kuju Entertainment	Midas Interactive Entertainment	
Ford Bold Moves Street Racing	2006	Razorworks	Eidos Interactive	
Ford Mustang: The Legend Lives	2005	Eutechnyx	2K Games	
Ford Racing 2	2003	Razorworks	2K Games	

Ford Racing 3	2004	Razorworks	2K Games	
Ford Street Racing	2006	Razorworks	Empire Interactive	
Forgotten Realms: Demon Stone	2004	Stormfront Studios	Atari	
Formula One 2001	2001	SCE Studio Liverpool	Sony Computer Entertainment	JP NA EU
Formula One 2002	2002	SCE Studio Liverpool	Sony Computer Entertainment	JP EU
Formula One 2003	2003	SCE Studio Liverpool	Sony Computer Entertainment	EU
Formula One 04	2004	SCE Studio Liverpool	Sony Computer Entertainment	JP EU
Formula One 05	2005	SCE Studio Liverpool	Sony Computer Entertainment	JP EU
Formula One 06	2006	SCE Studio Liverpool	Sony Computer Entertainment	JP EU
Franklin: A Birthday Surprise	2006	Neko Entertainment	The Game Factory	
Freakout	2001	Treasure	Conspiracy Entertainment	
Freaky Flyers	2003	Midway Games	Midway Games	
Freaky Sex Club				
Free Running	2007	Rebellion, Core Design	UbiSoft, Eidos Interactive	NA EU AUS
Freedom Fighters	2003	IO Interactive	Electronic Arts	
Freekstyle	2002	Electronic Arts, Page 44 Studios	Electronic Arts	
Frequency	2001	Harmonix Music Systems	Harmonix Music Systems	
Frogger: The Great Quest	2001	Konami	Konami	
Front Mission 4	2003	Square Enix	Square Enix	
Front Mission 5	2005	Square Enix	Square Enix	
Front Mission Online	2002	Square Enix	Square Enix	
Fugitive Hunter: War on Terror	2003	Black Ops	Encore Software	
Fullmetal Alchemist and the Broken Angel	2003	Racjin	Square Enix	
Fullmetal Alchemist 2: Curse of the Crimson Elixir	2004	Racjin	Square Enix	
Full Spectrum Warrior: Ten Hammers	2006	Pandemic Studios	THQ	JP NA EU AUS
Funkmaster Flex Digital Hitz Factory	2004	Jester Interactive	XS Games	
Fur Fighters: Viggo's Revenge	2001	Bizarre Creations	Acclaim Entertainment	
Futurama	2003	Unique Development Studios	VU Games, Fox Interactive	
Future Tactics: The Uprising	2004	Zed Two	Crave Entertainment, JoWood Productions	

G

Title and source	Year	Developer(s)	Publisher(s)	Regions released
G Surfers	2002	Blade Interactive	Midas Interactive Entertainment	
Gadget Racers	2001	Barnhouse Effect	Conspiracy Entertainment	
Gaelic Games: Football	2005	IR Gurus	Sony Computer Entertainment	EU
Gaelic Games: Football 2	2007	IR Gurus	Sony Computer Entertainment	EU

Gaelic Games: Hurling	2007	IR Gurus	Sony Computer Entertainment	EU
Galactic Wrestling: Featuring Ultimate Muscle	2004			
Galerians: Ash	2002	Polygon Magic	Sammy Studios/Enterbrain	JP NA EU AUS
Galidor: Defenders of the Outer Dimension		Electronic Arts	Asylum Entertainment	
Galleon: Islands of Mystery		Confounding Factor	SCi Entertainment	
Gallop Racer 2004	2004	Tecmo	Tecmo	
Gallop Racer 2006	2006	Tecmo	Tecmo	
Garfield: Saving Arlene		Hip Interactive	The Game Factory	
Garou: Mark of the Wolves	2005	SNK Playmore	SNK Playmore	JP
Gauntlet: Dark Legacy	2001	Midway Games West	Midway Games	JP NA EU
Gauntlet: Seven Sorrows	2005	Midway Games	Midway Games	NA EU AUS
Genji: Dawn of the Samurai	2005	Game Republic	Sony Computer Entertainment	JP NA EU AUS
The Getaway	2002	Team SOHO	Sony Computer Entertainment	JP NA EU
The Getaway: Black Monday	2004	Team SOHO	Sony Computer Entertainment	JP NA EU
Get On Da Mic	2004	A2M/Highway 1 Productions	Eidos Interactive	NA
Ghostbusters: The Video Game	2009	Red Fly Studio	Atari	NA EU
Ghosthunter	2003	SCE Studio Cambridge	Sony Computer Entertainment/Namco	JP NA EU
Ghost in the Shell: Stand Alone Complex	2004	Cavia	Bandai	JP NA EU AUS
Ghost Master	2003	Sick Puppies	Empire Interactive	
Ghost Rider	2007	Climax Group	2K Games	NA EU AUS
Giants: Citizen Kabuto	2001	Planet Moon Studios, Digital Mayhem	Interplay Entertainment	
Gitaroo Man	2001	Koei	Koei	
Gladiator: Sword of Vengeance	2003	Acclaim Studios Manchester	Acclaim Entertainment	
Gladius	2003	LucasArts	LucasArts/Activision	
Glass Rose	2003	Cing	Capcom	JP EU
Go Go Copter				
Go Go Golf				EU
Goblin Commander: Unleash the Horde	2003	Jaleco	Jaleco	NA
God Hand	2007	Clover Studio	Capcom	
God of War	2005	SCE Studios Santa Monica	Sony Computer Entertainment	JP NA EU
God of War II	2007	SCE Studios Santa Monica	Sony Computer Entertainment	JP NA EU AUS
Godai Elemental Force	2002	3DO	3DO	NA EU
The Godfather: The Game	2006	Electronic Arts	Electronic Arts	JP NA EU AUS
Godzilla: Save the Earth	2004	Pipeworks Software	Atari	
Godzilla: Unleashed	2007	Pipeworks Software	Atari	

Golden Age of Racing	2005	Midas Interactive Entertainment	Midas Interactive Entertainment	
The Golden Compass (video game)	2007	Shiny Entertainment	Sega	
GoldenEye: Rogue Agent	2004	Electronic Arts	Electronic Arts	JP NA EU AUS
Gradius III and IV	2000	Konami	Konami	
Gradius V	2004	Konami	Konami	
Graffiti Kingdom	2004	Taito/Garakuta Studio	Taito	
Gran Turismo Concept 2001 Tokyo	2002	Polyphony Digital	Sony Computer Entertainment	
Gran Turismo Concept 2002 Tokyo-Geneva	2002	Polyphony Digital	Sony Computer Entertainment	
Gran Turismo 3 A-spec	2001	Polyphony Digital	Sony Computer Entertainment	JP NA EU
Gran Turismo 4 Prologue	2003	Polyphony Digital	Sony Computer Entertainment	JP EU
Gran Turismo 4	2004	Polyphony Digital	Sony Computer Entertainment	JP NA EU
Grand Prix Challenge	2002	Melbourne House	Atari	JP NA EU AUS
Grand Slam Tennis	2009	EA Canada	EA Sports	NA UK INT AUS
Grand Theft Auto III	2001	DMA Design	Rockstar Games Capcom (JP)	JP NA EU AUS
Grand Theft Auto: Liberty City Stories	2005	Rockstar North, Rockstar Leeds	Rockstar Games Capcom (JP)	JP NA EU AUS
Grand Theft Auto: San Andreas	2004	Rockstar North	Rockstar Games Capcom (JP)	JP NA EU AUS
Grand Theft Auto: Vice City	2002	Rockstar North	Rockstar Games Capcom (JP)	JP NA EU AUS
Grand Theft Auto: Vice City Stories	2006	Rockstar North, Rockstar Leeds	Rockstar Games Capcom (JP)	JP NA EU AUS
Grandia II	2002	Game Arts	Enix/Ubisoft	
Grandia III	2005	Game Arts	Square Enix	
Grandia Xtreme	2002	Game Arts	Enix	
Gravity Games Bike: Street. Vert. Dirt.	2002	Midway Games	Midway Games	
Greg Hastings' Tournament Paintball Max'd	2005	The Whole Experience	Activision	
Gregory Horror Show	2003	Capcom	Capcom	
The Grim Adventures of Billy & Mandy	2006	High Voltage Software	Midway Games	
Growlanser Generations	2003	Career Soft	Atlus/Working Designs	
GT-R 400	2004	Kuju Entertainment	Midas Interactive	
GT-R Touring	2006	Kuju Entertainment	Midas Interactive	
Guilty Gear Isuka	2004	Arc System Works	Sammy Corporation/505 Game Street	
Guilty Gear X	2002	Arc System Works	Sammy Corporation	
Guilty Gear XX: Accent Core	2007	Arc System Works	Sammy Studios	
Guilty Gear XX: Accent Core Plus	2008	Arc System Works	Arc System Works/Aksys Games	
Guilty Gear XX: The Midnight Carnival	2002	Arc System Works	Sammy Studios	

Guilty Gear XX #reload	2003	Arc System Works	Sammy Studios	
Guilty Gear XX: Slash	2006	Arc System Works	Sammy Studios/Sega	
GuitarFreaks 3rd Mix & DrumMania 2nd Mix	2000	Konami Computer Entertainment Japan	Konami	JP
GuitarFreaks 4th Mix & DrumMania 3rd Mix (Gitadora!)	2001	Konami Computer Entertainment Japan	Konami	JP
GuitarFreaks V & DrumMania V	2006	Konami Digital Entertainment	Konami Digital Entertainment	JP
GuitarFreaks V2 & DrumMania V2	2006	Konami Digital Entertainment	Konami Digital Entertainment	JP
GuitarFreaks V3 & DrumMania V3	2007	Konami Digital Entertainment	Konami Digital Entertainment	JP
GuitarFreaks & DrumMania: Masterpiece Gold	2007	Konami Digital Entertainment	Konami Digital Entertainment	JP
GuitarFreaks & DrumMania: Masterpiece Silver	2006	Konami Digital Entertainment	Konami Digital Entertainment	JP
Guitar Hero	2005	RedOctane	Harmonix/MTV Games	
Guitar Hero II	2006	RedOctane	Activision/Harmonix	
Guitar Hero Encore: Rocks the 80s	2007	RedOctane	Activision/Harmonix	
Guitar Hero III: Legends of Rock	2007	RedOctane/Neversoft/Budcat Creations	Activision	
Guitar Hero: Aerosmith	2008	RedOctane/Neversoft/Budcat Creations	Activision	
Guitar Hero: Metallica	2009	Budcat Creations	Activision	
Guitar Hero World Tour	2008	RedOctane/Neversoft/Budcat Creations	Activision	
Gumball 3000	2002	Climax Studios	SCi Entertainment	
GUN	2005	Neversoft	Activision	NA EU
Gungriffon Blaze	2000	Game Arts	Working Designs	NA
Gunfighter 2	2003	Rebellion Developments	Ubisoft	
Gungrave	2002	Red Entertainment	Sega	
Gungrave: Overdose	2004	Ikusabune	Red Entertainment/Mastiff	
Gunslinger Girl Vol.1				
Gunslinger Girl Vol.2				
Gunslingter Girl Vol.3				
The Guy Game	2004	Top Heavy Studios	Gathering	

H

Title and source	Year	Developer(s)	Publisher(s)	Regions released
Half-Life (remake)	2001	*Valve Software*	*Sierra Entertainment*	NA
Hard Hitter Tennis				
Hard Rock Casino	2006	FarSight Studios	Crave Entertainment	NA
Hardware: Online Arena				

Harry Potter and the Chamber of Secrets	2002	warthorg	warthog	
Harry Potter and the Goblet of Fire	2005	Electronic Arts	Electronic Arts	
Harry Potter and the Half-Blood Prince	2009	Electronic Arts	Electronic Arts	
Harry Potter and the Order of the Phoenix	2007	Electronic Arts	Electronic Arts	
Harry Potter and the Philosopher's Stone	2003	Warthog	Electronic Arts	
Harry Potter and the Prisoner of Azkaban	2004	Electronic Arts and warthog	Electronic Arts and warthog	
Harry Potter: Quidditch World Cup	2004	Electronic Arts		
Harvest Moon: Save the Homeland	2001	Victor Interactive Software	Natsume	JP NA
Harvest Moon: A Wonderful Life		Natsume		JP NA
Haunting Ground				
Haven: Call of the King				
Hawk-Kawasaki Racing				
Headhunter				
He-Man Masters of the Universe-Defender of Greyskull				
Hello Kitty: Roller Rescue				
Heracles: Battle with the Gods				
Herdy Gerdy				
Heroes of Might and Magic: Quest for the Dragon Bone Staff	2001	New World Computing	3DO	NA
Heroes of the Pacific				
Hidden Invasion				
High School Musical: Sing It!				
High Rollers Casino	2004	Mud Duck Productions	ZeniMax Media	NA
Hi Hi Puffy AmiYumi 2				
Hitman: Blood Money				
Hitman: Contracts				
Hitman 2: Silent Assassin				
Hoihoi-san				
Homura				
Hot Shots Golf 3	2002	Clap Hanz	SCEI	NA
Hot Shots Golf Fore!	2003	Clap Hanz	SCEI	JP NA EU
Hot Wheels Velocity X				
Hot Wheels World Race				
HSX HyperSonic.Xtreme		Majesco		
The Hulk				
Hummer Badlands				
Hunter: The Reckoning: Wayward				
The Hustle: Detroit Streets				
Hype Time Quest				

I

Title and source	Year	Developer(s)	Publisher(s)	Regions released
Ico	2001	Team Ico	SCEI	JP NA EU
I-Ninja				
Ice Age 2: The Meltdown				
IGPX				
IHRA Drag Racing 2				
IHRA Drag Racing: Sportsman Edition				
Ikusagami				
The Incredible Hulk: Ultimate Destruction				
The Incredibles: Rise of the Underminer				
Indiana Jones and the Emperor's Tomb	2003	The Collective, Inc.	LucasArts	
Indiana Jones and the Staff of Kings	2009	LucasArts	LucasArts	
IndyCar Series	2003	Brain in a Jar	Codemasters	NA EU
IndyCar Series 2005	2004	Codemasters	Codemasters	NA EU
Initial D: Special Stage		Sega		
Inspector Gadget				
Intellivision Lives!				
International Cue Club				
International Cue Club 2				
International Golf Pro				
International Super Karts				
International Superstar Soccer				
Interview With a Made Man				
Ikki Tousen: Shining Dragon				
In The Groove				
InuYasha: Feudal Combat		Bandai		
InuYasha: The Secret of the Cursed Mask		Bandai		
Island Xtreme Stunts				
The Italian Job				

J

Title and source	Year	Developer(s)	Publisher(s)	Regions released
J-League Winning Eleven 6	2002	Konami	Konami	JP
Jackass: The Game	2007	Sidhe Interactive	Red Mile Entertainment	
Jackie Chan Adventures	2005	Atomic Planet	Hip Interactive, SCEI	
Jacked	2005	Sproing	Empire Interactive, 3DO	

Jade Cocoon 2	2001	Genki	Ubisoft	
Jak and Daxter: The Precursor Legacy	2001	Naughty Dog Software	SCEI	
Jak II	2003	Naughty Dog Software	SCEI	
Jak 3	2004	Naughty Dog Software	SCEI	
Jak X: Combat Racing	2005	Naughty Dog Software	SCEI	
Jak and Daxter: The Lost Frontier	2009	High Impact Games	SCEI	
James Bond 007: Agent Under Fire	2001	Electronic Arts	Electronic Arts	
James Bond 007: Everything or Nothing	2004	Electronic Arts	Electronic Arts	
James Bond 007: From Russia with Love	2005	Electronic Arts	Electronic Arts	
James Bond 007: Nightfire	2002	Electronic Arts	Electronic Arts	
James Bond 007: Quantum of Solace	2008	Eurocom	Activision	
James Cameron's Dark Angel	2002	Radical Entertainment	Sierra	
Jeep Thrills	2008	Game Sauce	DSI Games	NA, EU
Jeopardy!	2003	Artech Studios	Infogrames	
Juiced	2005	Juice Games	THQ	
Juiced 2	2007	Juice Games	THQ	
Judge Dredd: Dredd vs Death	2005	Rebellion	Evolved Games	
Jurassic Park: Operation Genesis	2003	Blue Tongue	Universal Interactive, Konami	
Just Cause	2006	Avalanche Studios	Eidos Interactive	

K

Title and source	Year	Developer(s)	Publisher(s)	Regions released
Kanon	2002	Key	NEC Interchannel	JP
Kao the Kangaroo Round 2	2005	Tate Interactive	Atari, JoWood Productions	
Katamari Damacy	2004	Namco	Namco	JP NA
Kengo Master of Bushido	2001	Crave	Ubisoft	
Kessen	2000	Koei	Koei	JP NA EU
Killer7	2005	*Grasshopper Manufacture*	*Capcom*	JP NA EU
*Killzone (**Guerilla Games**)*	2004	*Guerrilla Games*	*Sony Computer Entertainment Europe*	
Kinetica				
King Arthur		Konami		
Kingdom Hearts	2002	Squaresoft	Squaresoft	JP NA EU
Kingdom Hearts Final Mix	2003	Squaresoft	Squaresoft	JP
Kingdom Hearts II	2006	Square Enix	Square Enix	JP NA EU
Kingdom Hearts II Final Mix +	2007	Square Enix	Square Enix	JP
Kingdom Hearts Re: Chain of Memories	2008	Square Enix	Square Enix	JP NA

Kingdom Hearts III	TBD	Square Enix	Square Enix	JP
King of Clubs		Oxygen Interactive		
The King of Fighters '94 Re-bout		SNK Playmore	SNK Playmore	JP
The King of Fighters '98 Ultimate Match	2008(JP) 2009(NA)	SNK Playmore	SNK Playmore (JP) Ignition Entertainment (NA)	JP NA
The King of Fighters 2000		SNK Playmore	SNK Playmore	JP
The King of Fighters 2001		Eolith	SNK Playmore	JP
The King of Fighters 2000/2001		SNK Playmore	SNK Playmore	NA
The King of Fighters 2002		Eolith	SNK Playmore	JP
The King of Fighters 2002 Unlimited Match	2009	SNK Playmore	SNK Playmore	JP
The King of Fighters 2003		SNK Playmore	SNK Playmore	JP
The King of Fighters 2002/2003		SNK Playmore	SNK Playmore	NA
The King of Fighters Neowave		SNK Playmore	SNK Playmore	JP NA
The King of Fighters XI		SNK Playmore	SNK Playmore	JP NA
The King of Fighters Collection: The Orochi Saga		SNK Playmore	SNK Playmore	NA
King's Field: The Ancient City	2001	FromSoftware	Agetec	JP NA
Kira Kira Rock N Roll Show		Overdrive	Princess Soft	JP
Klonoa 2: Lunatea's Veil				
Knight Rider: The Game		Davilex Games		
Knight Rider 2 : The Game		Davilex Games		
Knights of the Temple: Infernal Crusade	2004	Starbreeze Studios	TDK Mediactive	EU
Knockout Kings 2002				
Knockout Kings 2003				
KOF: Maximum Impact		SNK Playmore	SNK Playmore	JP NA
KOF: Maximum Impact 2 (JP) *The King of Fighters 2006* (NA)		SNK Playmore	SNK Playmore	JP NA
KOF: Maximum Impact Regulation A		SNK Playmore	SNK Playmore	JP
Konjiki no Gash Bell: Yuujou no Tag Battle		8ing	Bandai	
Kujibiki Unbalance: Kaicho Onegai Smash Fight				
Kuon				
Kuri Kuri Mix				
Kya: Dark Lineage				

L

Title and source	Year	Developer(s)	Publisher(s)	Regions released
L.A. Rush	2005	Midway Games	Midway Games	NA EU
La Pucelle: Tactics	2002	Nippon Ichi	Nippon Ichi, Mastiff	JP NA EU AUS
LarryBoy and the Bad Apple	2006	Papaya Studios	Crave Entertainment	NA
Leaderboard Golf	2006	Aqua Pacific	Midas Interactive Entertainment	EU
League Series Baseball 2	2001	Mahou	Midas Interactive Entertainment	JP EU
Legacy of Kain: Defiance	2003	Crystal Dynamics	Eidos Interactive	NA EU
Legacy of Kain: Soul Reaver 2	2001	Crystal Dynamics	Eidos Interactive	JP NA EU
Legaia 2: Duel Saga	2002	Prokion	Eidos Interactive	JP NA EU
The Legend of Alon D'ar	2001	Stormfront Studios	Ubisoft	NA EU
Legend of Kay	2005	Neon Studios	Capcom	NA EU AUS
Legend of the Dragon	2007	Neko Entertainment	The Game Factory	NA
The Legend of Spyro: A New Beginning	2006	Krome Studios	Sierra Entertainment, Activision Blizzard	NA EU AUS
The Legend of Spyro: The Eternal Night	2007	Krome Studios	Sierra Entertainment	NA EU AUS
The Legend of Spyro: Dawn of the Dragon	2008	Etranges Libellules	Sierra Entertainment, Activision Blizzard	NA EU AUS
Legends of Wrestling	2001	Acclaim Entertainment	Acclaim Entertainment	NA EU AUS
Legends of Wrestling II	2002	Acclaim Entertainment	Acclaim Entertainment	NA EU AUS
Lego Batman: The Video Game	2008	Traveller's Tales	Warner Bros. Interactive Entertainment	NA EU AUS
Lego Indiana Jones: The Original Adventures	2008	Traveller's Tales	LucasArts	NA EU AUS
Lego Star Wars: The Video Game	2005	Traveller's Tales	Eidos Interactive, LucasArts	JP NA EU AUS
Lego Star Wars II: The Original Trilogy	2006	Traveller's Tales, Amaze Entertainment	LucasArts, TT Games	JP NA EU AUS
Leisure Suit Larry: Magna Cum Laude	2004	High Voltage Software	Sierra Entertainment	NA
Lemmings	2006	Team17, Rusty Nutz	SCEI	EU AUS
Lethal Skies Elite Pilot: Team SW	2001	Asmik Ace Enterainment, Bit Town	Sammy Corporation	JP NA
Let's Make a Soccer Team!	2006	Sega Europe	Sega	EU AUS
Let's Ride: Silver Bucket Stables	2006	Coresoft	ValuSoft	NA
Lifeline	2003	SCEI	SCEI, Konami	JP NA
Little Britain: The Video Game	2007	Revolution Studios	Blast! Entertainment	EU
Little Busters! Converted Edition	2009	Key	Prototype	JP
LMA Manager 2002	2001	Codemasters	Codemasters	EU
LMA Manager 2003	2002	Codemasters		JP EU AUS
LMA Manager 2004	2004	Codemasters	Codemasters	JP EU AUS
LMA Manager 2005	2005	Codemasters	Codemasters	JP EU AUS
LMA Manager 2006	2005	Codemasters	Codemasters	JP EU AUS

LMA Manager 2007	2006	Codemasters	Codemasters	JP EU AUS
London Racer 2	2002	Davilex Games	Davilex Games	EU
London Racer World Challenge	2003	Davilex Games	Koch Media	EU
Looney Tunes: Acme Arsenal	2007	Redtribe	Warner Bros.	JP NA EU AUS
Looney Tunes: Back in Action	2003	Warthog	Electronic Arts, Warner Bros.	NA
Looney Tunes: Space Race	2002	Infogrames Melbourne	Infogrames	NA EU
The Lord of the Rings: The Fellowship of the Ring	2002	Surreal Software	Vivendi Universal	NA EU
The Lord of the Rings: The Two Towers	2002	Stormfront Studios	Electronic Arts	JP NA EU AUS
The Lord of the Rings: The Return of the King	2003	Electronic Arts	Electronic Arts	JP NA EU AUS
The Lord of the Rings: The Third Age	2004	Electronic Arts	Electronic Arts	JP NA EU AUS
Lowrider	2002	Jaleco	Jaleco	JP NA
Lumines Plus	2007	Q Entertainment	Disney Interactive Studios	NA
Lupin the 3rd: Treasure of the Sorcerer King	2002	Banpresto	Bandai	JP NA

M

Title and source	Year	Developer(s)	Publisher(s)	Regions released
Mace Griffin: Bounty Hunter	2004	Warthog Games	Vivendi	
Mad Maestro!	2001	Desert Productions	Eidos Interactive	JP NA EU AUS
Madagascar	2005	Toys for Bob	Activision	JP NA EU AUS
Madagascar: Escape 2 Africa	2008	Idolminds	Activision	
Madden NFL 2001	2000	EA Tiburon	EA Sports	NA
Madden NFL 2002	2001	EA Tiburon	EA Sports	JP NA EU AUS
Madden NFL 2003	2002	EA Tiburon	EA Sports	JP NA EU AUS
Madden NFL 2004	2003	EA Tiburon	EA Sports	JP NA
Madden NFL 2005	2004	EA Tiburon	EA Sports	JP NA EU AUS
Madden NFL 06	2005	EA Tiburon	EA Sports	JP NA EU AUS
Madden NFL 07	2006	EA Tiburon	EA Sports	JP NA EU AUS
Madden NFL 08	2007	EA Tiburon	EA Sports	JP NA EU AUS
Madden NFL 09	2008	EA Tiburon	EA Sports	JP NA EU AUS
Madden NFL 10	2009	EA Tiburon	EA Sports	JP NA EU AUS
Mafia	2002	Illusion Softworks	Gathering of Developers	
Magic Pengel	2003	Garakuda-Studio, Taito	Agetec	
Magix Music Maker				
Magister Negi Magi: 1st time			Konami	
Magister Negi Magi: 2nd time			Konami	
Magna Carta: Tears of Blood	2004	Softmax	Banpresto, Atlus	JP NA EU AUS
Mahou Sensei Negima! 2-Jikanme				

Mai-Hime: The Another				
Major League Baseball 2K5	2005	Kush Games	2K Sports	
Major League Baseball 2K5: World Series Edition	2005	Kush Games	2K Sports	
Major League Baseball 2K6	2006	Visual Concepts	2K Sports	JP NA
Major League Baseball 2K7	2007	Kush Games	2K Sports	JP NA EU AUS
Major League Baseball 2K8	2008	Kush Games	2K Sports	JP NA EU
Major League Baseball 2K9	2009	Visual Concepts	2K Sports	NA
Makai Kingdom: Chronicles Of The Sacred Tome	2005	Nippon Ichi	Nippon Ichi, KOEI	JP NA EU
Maken Shao	2001	Atlus	Midas Interactive Entertainment	JP EU AUS
Malice	2004	Argonaut Games	Mud Duck Games, Evolved Games	NA EU
Manchester United Club Football 2005	2004	Codemasters	Codemasters	
Manhunt	2003	Rockstar North	Rockstar Games	NA EU
Manhunt 2	2007	Rockstar Games	Rockstar Games	JP NA EU AUS
Marc Ecko's Getting Up: Contents Under Pressure	2006	The Collective, Inc.	Atari	NA EU
Mar Heaven: Arm Fight				
Mark Davis Pro Bass Challenge	2003	SIMS	Natsume	NA EU
The Mark of Kri	2003	Sony Computer Entertainment	Sony Computer Entertainment	JP NA EU
Martial Arts: Capoeira	2009	Twelve Interactive	Graffiti Entertainment	
Marvel Nemesis	2005	Nihilistic Software	Electronic Arts	JP NA EU
Marvel vs. Capcom 2	2002	Capcom	Capcom	
Marvel: Ultimate Alliance	2006	Raven Software	Activision	
Marvel Ultimate Alliance 2: Fusion	2009	Vicarious Visions	Activision	
Mary-Kate and Ashley: Sweet 16	2002	n-Space	Acclaim Entertainment	
Mashed	2004	Supersonic Software	Empire Interactive	
Mashed: Drive To Survive				
Mashed Fully Loaded				
Master Rallye				
Mat Hoffman's Pro BMX 2	2002	Rainbow Studios	Activision	
The Matrix: Path of Neo	2005	Shiny Entertainment	Atari	NA EU
Maximo vs. Army of Zin	2003	Capcom	Capcom	
Maximo: Ghosts to Glory	2001	Capcom	Capcom	
Max Payne	2001	Rockstar Games	Rockstar Games	
Max Payne 2: The Fall of Max Payne	2003	Rockstar Games	Rockstar Games	
MaXXed Out Racing				
Maze Action				
McFarlane's Evil Prophecy	2004	Konami	Konami	
MDK2: Armageddon	2001	BioWare	Interplay Entertainment	

Medal of Honor: European Assault	2005	EA	EA	
Medal of Honor: Frontline	2002	EA	EA	
Medal of Honor: Rising Sun	2003	EA	EA	
Medal of Honor: Vanguard	2007	EA	EA	
Mega Man Anniversary Collection	2004	Atomic Planet Entertainment	Capcom	
Mega Man X Collection	2006	Capcom	Capcom	
Mega Man X Command Mission	2004	Capcom	Capcom	
Mega Man X7	2003	Capcom	Capcom	
Mega Man X8	2004	Capcom	Capcom	
MegaRace 3: Nanotech Disaster	2001			
Melty Blood: Act Cadenza	2006		Ecole Software	JP
Melty Blood: Actress Again	2009		Ecole Software	JP
Men in Black II: Alien Escape	2002	Melbourne House	Infogrames	
Mercedes-Benz World Racing	2003	Synetic	TDK Mediactive	
Mercenaries: Playground of Destruction	2005	Pandemic Studios	LucasArts	JP NA EU AUS
Mercenaries 2: World in Flames	2008	Pi Studios	Electronic Arts	
Mercury Meltdown Remix	2006	Ignition Entertainment	Ignition Entertainment, SCEI, Atari	JP NA EU
Metal Arms: Glitch in the System	2003	Swingin' Ape Studios	Vivendi Universal	
Metal Gear Solid 2: Sons of Liberty	2001	Konami	Konami	JP NA EU SK
Metal Gear Solid 2: Substance	2003	Konami	Konami	JP NA EU
Metal Gear Solid 3: Snake Eater	2004	Konami	Konami	JP NA EU KOR AUS
Metal Gear Solid 3: Subsistence	2005	Konami	Konami	JP NA EU
Metal Saga	2005	Createch	Success, Atlus Software	
Metal Slug 3	2003	SNK	SNK	
Metal Slug 4	2002	Mega Enterprise, BrezzaSoft	SNK Playmore	
Metal Slug 4 And 5	2005	BrezzaSoft	SNK Playmore	
Metal Slug 6	2006	SNK Playmore	SNK Playmore	
Metropolismania	2002	Media Factory	Natsume	
Micro Machines V4	2006	Supersonic Software	Codemasters	
Midnight Club Street Racing	2000	Angel Studios	Rockstar Games	
Midnight Club 2	2003	Rockstar Games	Rockstar Games	
Midnight Club 3: DUB Edition	2005	Rockstar Games	Rockstar Games	
Midnight Club 3: DUB Edition Remix	2006	Rockstar Games	Rockstar Games	
Midway Arcade Treasures	2003	Digital Eclipse	Midway Games	
Midway Arcade Treasures 2	2004	Digital Eclipse	Midway Games	
Midway Arcade Treasures 3	2005	Midway Games	Midway Games	
Mike Tyson Heavyweight Boxing	2002	Atomic Planet	Codemasters	

Mini Desktop Racing	2005	Data Design Interactive	Metro3D	
Mission Impossible: Operation Surma	2003	Paradigm Entertainment	Atari	
Mister Mosquito	2001	Zoom	Sony Computer Entertainment, Eidos Interactive	
MLB 2004	2003	989 Sports	Sony Computer Entertainment	
MLB 2005	2004	989 Sports	Sony Computer Entertainment	
MLB 2006	2005	989 Sports	Sony Computer Entertainment	
MLB '06: The Show	2006	Sony Computer Entertainment	Sony Computer Entertainment	NA
MLB '07: The Show	2007	Sony Computer Entertainment	Sony Computer Entertainment	NA
MLB '08: The Show	2008	Sony Computer Entertainment	Sony Computer Entertainment	NA
MLB '09: The Show	2009	Sony Computer Entertainment	Sony Computer Entertainment	NA
MLB Power Pros	2007	Konami	2K Sports	
MLB Power Pros 2008	2008	Konami	2K Sports	JP NA
MLB SlugFest 2003	2002	Midway Games	Midway Games	
MLB SlugFest 2004	2003	Midway Games	Midway Games	
MLB SlugFest 2006	2005	Midway Games	Midway Games	
MLB SlugFest: Loaded	2004	Midway Games	Midway Games	
Mobile Light Force 2	2003	Alfa System	XS Games	
Mobile Suit Gundam SEED: Never Ending Tomorrow	2005	Banpresto	Bandai	
Mobile Suit Gundam SEED Destiny: Generation Of C.E.	2005	Bandai	Bandai	
Mobile Suit Gundam: AEUG v.s Titans	2003	Capcom	Bandai	
Mobile Suit Gundam: Climax U.C.	2003	Bandai	Bandai	
Mobile Suit Gundam: Encounters in Space	2003	Bandai	Bandai	
Mobile Suit Gundam: Journey to Jaburo	2001	Bandai	Bandai	
Mobile Suit Gundam: The One Year War	2005	Namco	Bandai	
Mobile Suit Gundam: Gundam vs. Zeta Gundam	2004	Bandai	Bandai	
Mobile Suit Gundam: Zeonic Front	2001	Sunrise Interactive	Bandai	
Moderngroove: Ministry of Sound Edition	2001	Moderngroove Entertainment	Ubisoft	
Mojib-Ribbon	2003	NanaOn-Sha	Sony Computer Entertainment	
Mojo!	2003	FarSight Studios	Crave Entertainment	
Monopoly Party	2002	Runecraft	Infogrames	
Monster 4x4: Masters of Metal	2003	Ubisoft	Ubisoft	
Monster House	2006	A2M	THQ	
Monster Hunter	2004	Capcom	Capcom	JP NA EU

Monster Hunter G	2005	Capcom	Capcom	JP
Monster Jam: Maximum Destruction	2002	Unique Development Studios	Ubisoft	
Monster Rancher 3	2001	Tecmo	Tecmo	
Monster Rancher 4	2003	Tecmo	Tecmo	
Monster Rancher EVO	2005	Tecmo	Tecmo	
Monsters, Inc. Scare Island	2002			
Monsters vs. Aliens	2009	Activision	Activision	
Mortal Kombat: Armageddon	2006	Midway Games	Midway Games	
Mortal Kombat: Deadly Alliance	2002	Midway Games	Midway Games	
Mortal Kombat: Deception	2004	Midway Games	Midway Games	
Mortal Kombat: Shaolin Monks	2005	Midway Games	Midway Games	
Motocross Mania 3	2005	Deibus Studios	2K Games	
MotoGP	2000	Namco	Namco	
MotoGP 2	2002	Namco	Namco	
MotoGP 3	2003	Namco	Namco	
MotoGP 4	2006	Namco	Namco	
MotoGP '07	2007	Milestone	Capcom	JP NA EU AUS
MotoGP '08	2008	Milestone	Capcom	
Motor Mayhem				
Mr Golf				
Ms. Chris Grams LA game				
MS Saga: A New Dawn	2005	Bandai	Bandai	
MTV Music Generator 2				
MTV Music Generator 3: This Is The Remix				
MTX: Mototrax	2004	Left Field Productions	Activision	
The Mummy Returns	2001	Blitz Games	Vivendi Universal Games	NA EU AUS
The Mummy: Tomb of the Dragon Emperor	2008	Eurocom	Sierra Entertainment	NA EU AUS
Muppets Party Cruise	2003	Mass Media	TDK Mediactive	
Musashi: Samurai Legend	2005	Square Enix	Square Enix	
Music 3000	2003	Jester Interactive	Jester Interactive	
MVP Baseball 2003	2003	EA Canada	Electronic Arts	
MVP Baseball 2004	2004	EA Canada	Electronic Arts	
MVP Baseball 2005	2005	EA Canada	Electronic Arts	
MVP 06: NCAA Baseball	2006	EA Canada	Electronic Arts	
MVP 07: NCAA Baseball	2007	EA Canada	Electronic Arts	
MX 2002 Featuring Ricky Carmichael	2001	Pacific Coast Power and Light	THQ	
MX Super Fly	2002	Pacific Coast Power and Light	THQ	

MX Unleashed	2004	Rainbow Studios	THQ	
MX vs. ATV Unleashed	2005	Rainbow Studios	THQ	
MX vs. ATV Untamed	2007	Incinerator Games	THQ	
MX World Tour: Featuring Jamie Little	2005	Impulse Games	Crave Entertainment	
MXRider	2001	Paradigm Entertainment	Atari, Infogrames	
My Street	2003	Idolminds	Sony Computer Entertainment	
My-Otome: Otome Butoushi!!		Sunrise Interactive		
Myst III: Exile	2002	Presto Studios	Ubisoft	
Mystic Heroes	2002	Koei	Koei	

N

Title and source	Year	Developer(s)	Publisher(s)	Regions released
Namco Museum	2001	Mass Media	Namco	
Namco Museum 50th Anniversary Arcade Collection	2005	Namco	Namco	
Namco X Capcom	2005	Monolith Soft	Namco	JP
Nano Breaker	2005	Konami	Konami	JP NA
NARC	2005	Midway	Midway	
Naruto: Ultimate Ninja	2003	Cyberconnect2		JP NA EU
Naruto: Ultimate Ninja 2	2004	Cyberconnect2		JP NA EU
Naruto: Ultimate Ninja 3	2005	Cyberconnect2		JP NA EU
Naruto: Uzumaki Ninden	2005			JP NA
Naruto: Uzumaki Chronicles 2	2007	Cavia	Namco Bandai Games	
Naruto Shippūden: Ultimate Ninja 4		Cyberconnect2		JP NA
Naruto Shippuden: Narutimate Accel 2	2007	Cyberconnect2		JP
NASCAR: Dirt to Daytona	2002	Monster Games	Infogrames	
NASCAR Heat 2002	2001	Monster Games	Infogrames	
NASCAR 2001	2000	EA Sports	EA Sports	
NASCAR Thunder 2002	2001	EA Tiburon	Electronic Arts	
NASCAR Thunder 2003	2002	EA Sports	Electronic Arts	
NASCAR Thunder 2004	2003	EA Tiburon	Electronic Arts	
NASCAR 2005: Chase for the Cup	2004	EA Tiburon	Electronic Arts	NA
NASCAR 06: Total Team Control	2005	EA Tiburon	Electronic Arts	NA EU
NASCAR 07 (**EA Sports**)	2006	EA Tiburon	Electronic Arts	NA
NASCAR 08	2007	EA Tiburon	EA Sports	NA EU AUS
NASCAR 09	2008	EA Tiburon	EA Sports	
Naval Ops: Commander	2004	Micro Cabin	Koei	JP NA EU
Naval Ops: Warship Gunner	2003	Micro Cabin	Koei	JP NA EU

Naval Ops: Warship Gunner 2	2006	Micro Cabin	Koei	JP NA EU
NBA 06	2005	SCE Studio San Diego, 989 Sports	Sony Computer Entertainment	
NBA 07	2006	SCE Studio San Diego	Sony Computer Entertainment	
NBA 08	2007	SCE Studio San Diego	Sony Computer Entertainment	
NBA '09: The Inside	2008	SCE Studio San Diego	Sony Computer Entertainment	
NBA 2K2	2001	Visual Concepts	Sega Sports	
NBA 2K3	2002	Visual Concepts	Sega Sports	
NBA 2K6	2005	Visual Concepts	2K Sports	
NBA 2K7	2006	Visual Concepts	2K Sports	
NBA 2K8	2007	Visual Concepts	2K Sports	
NBA 2K9	2008	Visual Concepts	2K Sports	
NBA 2K10	2009	Visual Concepts	2K Sports	
NBA Ballers	2004	Midway	Midway	NA
NBA Ballers Phenom	2006	Midway	Midway	
NBA Hoopz	2001	Midway	Midway	
NBA Jam	2003	Acclaim Entertainment	Acclaim Entertainment	
NBA Live 2001	2001	EA Canada	EA Sports	
NBA Live 2002	2001	EA Canada	EA Sports	
NBA Live 2003	2002	EA Canada	EA Sports	
NBA Live 2004	2003	EA Canada	EA Sports	
NBA Live 2005	2004	EA Canada	EA Sports	
NBA Live 06	2005	EA Canada	EA Sports	
NBA Live 07	2006	EA Canada	EA Sports	
NBA Live 08	2007	HB Studios	EA Sports	
NBA Live 09	2008	HB Studios	EA Sports	
NBA Live 10	2009	HB Studios	EA Sports	
NBA Shootout 2001	2001	989 Sports	Sony Computer Entertainment	
NBA Shootout 2003	2002	989 Sports	Sony Computer Entertainment	
NBA Shootout 2004	2003	989 Sports	Sony Computer Entertainment	
NBA Starting Five	2002	Konami	Konami	
NBA Street	2001	EA Sports BIG	EA Sports BIG	
NBA Street Vol. 2	2003	EA Canada	EA Sports BIG	
NBA Street V3	2005	EA Sports BIG	EA Sports BIG	
NCAA College Basketball 2K3	2002	Kush Games, Visual Concepts	Sega	
NCAA College Football 2K3	2002	Visual Concepts	Sega	
NCAA Final Four 2001	2000	989 Sports	Sony Computer Entertainment	
NCAA Final Four 2002	2001	989 Sports	Sony Computer Entertainment	
NCAA Final Four 2003	2002	989 Sports	Sony Computer Entertainment	

NCAA Final Four 2004	2003	989 Sports	Sony Computer Entertainment	
NCAA Football 2002	2001	Tiburon	Electronic Arts	
NCAA Football 2003	2002	Tiburon	Electronic Arts	
NCAA Football 2004	2003	Tiburon	Electronic Arts	
NCAA Football 2005	2004	EA Tiburon	Electronic Arts	
NCAA Football 06	2005	EA Tiburon	Electronic Arts	
NCAA Football 07	2006	EA Tiburon	Electronic Arts	
NCAA Football 08	2007	EA Tiburon	Electronic Arts	
NCAA Football 09	2008	EA Tiburon	Electronic Arts	
NCAA Gamebreaker 2001	2000	989 Sports	Sony Computer Entertainment	
NCAA Gamebreaker 2003	2002	989 Sports	Sony Computer Entertainment	
NCAA Gamebreaker 2004	2003	989 Sports	Sony Computer Entertainment	
NCAA March Madness 2002	2002	EA Canada	EA Sports	
NCAA March Madness 2003	2002	EA Canada	EA Sports	
NCAA March Madness 2004	2003	EA Canada	EA Sports	
NCAA March Madness 2005	2004	EA Canada	EA Sports	
NCAA March Madness 06	2005	EA Canada	EA Sports	
NCAA March Madness 07	2007	EA Canada	EA Sports	
NCAA March Madness 08	2008	EA Canada	EA Sports	
Need for Speed: Carbon	2006	EA Black Box	Electronic Arts	NA EU JP
Need For Speed: Hot Pursuit 2	2002	Black Box Games, EA Black Box	EA Games	
Need for Speed: Most Wanted	2005	EA Black Box	Electronic Arts	NA EU JP
Need for Speed: Most Wanted Black Edition				
Need for Speed ProStreet	2007	EA Black Box	Electronic Arts	NA EU JP
Need for Speed: Undercover	2008	EA Black Box	Electronic Arts	NA EU JP
Need for Speed: Underground	2003	EA Black Box	Electronic Arts	NA EU JP
Need for Speed: Underground 2	2004	EA Black Box	Electronic Arts	NA EU JP
Negima!?: 3rd time				
Negima!?: Dream Tactics		Konami		
Neo Contra	2004	Konami	Konami	JP NA EU
Neo Geo Battle Coliseum	2005	SNK Playmore	SNK Playmore (Japan & US), Ignition Entertainment (EU)	JP NA EU
Neopets: The Darkest Faerie	2005	Idol Minds Digital Entertainment	Sony Computer Entertainment	NA
Newcastle United Club Football 2005	2004	Codemasters	Codemasters	
Next Generation Tennis	2002	Wanadoo	Wanadoo	
Next Generation Tennis 2003	2003	Carapace	Wanadoo	
NFL 2K2	2001	Visual Concepts	Sega	
NFL 2K3	2002	Visual Concepts	Sega	

NFL Blitz 20-02	2001	Midway	Midway	
NFL Blitz 20-03	2002	Midway	Midway	
NFL Blitz Pro	2003	Midway	Midway	
NFL GameDay 2001	2000	Red Zone Entertainment	Sony Computer Entertainment	
NFL GameDay 2002	2001	Red Zone Entertainment	Sony Computer Entertainment	
NFL GameDay 2003	2002	Red Zone Entertainment	Sony Computer Entertainment	
NFL GameDay 2004	2003	Red Zone Entertainment	Sony Computer Entertainment	
NFL Head Coach	2006	EA Tiburon	EA Sports	
NFL Quarterback Club 2002	2001	Acclaim	Acclaim	
NFL Street	2004	EA Tiburon	EA Sports	
NFL Street 2	2004	EA Tiburon	EA Sports	
NFL Street 3	2006	EA Tiburon	EA Sports	
NHL 2001	2000	EA Canada	EA Sports	
NHL 2002	2001	EA Canada	EA Sports	
NHL 2003	2002	EA Canada	EA Sports	
NHL 2004	2003	EA Canada	EA Sports	
NHL 2005	2004	EA Canada	EA Sports	
NHL 06	2005	EA Canada	EA Sports	
NHL 07	2006	EA Montreal	EA Sports	
NHL 08	2007	HB Studios	EA Sports	
NHL 09	2008	HB Studios	EA Sports	
NHL 2K3	2002	Treyarch	Sega (US), Atari (EU)	
NHL 2K6	2005	Kush Games	2K Sports	
NHL 2K7	2006	Kush Games	2K Sports	
NHL 2K8	2007	Kush Games	2K Sports	
NHL 2K9	2008	Kush Games	2K Sports	
NHL FaceOff 2001	2001	989 Sports	Sony Computer Entertainment	
NHL FaceOff 2003	2002	989 Sports	Sony Computer Entertainment	
NHL Hitz 20-02	2001	Black Box Games	Midway Games	
NHL Hitz 20-03	2002	Black Box Games	Midway Games	
NHL Hitz Pro	2003	Next Level Games	Midway Games	
NHRA Drag Racing Countdown to the Championship 2007	2007	Pipeworks Software	THQ	
Nicktoons Movin'	2004	THQ	THQ	
Nicktoons Unite!	2005	Blue Tongue Entertainment	THQ	
Nicktoons: Battle for Volcano Island	2006	Blue Tongue Entertainment	THQ	
The Nightmare Before Christmas: Oogie's Revenge	2005	Capcom	Buena Vista Games	NA EU
The Nightmare of Druaga: Fushigino Dungeon	2004	Arika, Chunsoft	Namco	

Nightshade	2003	Wow Entertainment	Sega	
Ninja Assault	2002	Namco	Namco	JP NA EU
Nishanka's Ultimate Grand Turismo & Juiced				
Noble Racing	2006		Midas Interactive	
Nobunaga's Ambition: Rise to Power	2004	Koei	Koei	
Nobunaga's Ambition: Iron Triangle	2009	Koei	Koei	
NRA Gun Club	2006	Crave Entertainment	Crave Entertainment	
NTRA Breeders' Cup World Thoroughbred Championships	2005	4JStudios	Bethesda Softworks	

O

Title and source	Year	Developer(s)	Publisher(s)	Regions released
ObsCure	2004	Hydravision	DreamCatcher Games	NA EU
Obscure II				
Odin sphere	2007	Vanillaware	Atlus	JP NA EU
Okage: Shadow King	2001	Zener Works	SCE	
Ōkami	2006	Clover Studio	Capcom	JP NA EU
One Piece Grand Adventure				
One Piece Grand Battle				
One Piece: Pirates Carnival				
OneChanbara		Tamsoft		
Oni		Bungie Studios		
Onimusha: Warlords	2001	Capcom	Capcom	JP NA EU
Onimusha 2: Samurai's Destiny		Capcom		
Onimusha 3: Demon Siege		Capcom		
Onimusha Blade Warriors		Capcom		
Onimusha: Dawn of Dreams		Capcom		
Open Season				
Operation Air Assault				
The Operative: No One Lives Forever		Monolith		
Orphen: Scion of Sorcery				
Outlaw Golf				
Outlaw Golf 2				
'Outlaw Tennis				
Outlaw Volleyball Remixed				
OutRun 2006: Coast 2 Coast				
Over the Hedge				

P

Title and source	Year	Developer(s)	Publisher(s)	Regions released
Pacific Warriors II: Dogfight				
Pac-Man Fever		Mass Media	Namco	
Pac-Man World 2		Namco		
Pac-Man World 3				
Pac-Man World Rally				
PaRappa the Rapper 2		Sony		
Paris Dakar Rally				
Pasty Maker Sandwich Van Pay Out				
Penny Racers				
Perfect Ace Pro Tournament Tennis				
Peter Jackson's King Kong: The Official Game of the Movie		Ubisoft		
Phantom Brave		Nippon Ichi		
Phantasy Star Universe		Sega		
Phantom Crash 2050				
Pilot Down				
Pimp My Ride: Street Racing	2009	Activision	Activision	
Pinball Fun				
Pinball Hall of Fame: The Gottlieb Collection	2004	FarSight Studios	Crave Entertainment	NA
Pipo Saru 2001	2001	SCEI	SCEI	
Pirates: Legend of the Black Buccaneer				
Pirates: The Legend of Black Kat				
Pirates of the Caribbean: At Worlds End				
Pirates of the Caribbean: The Legend of Jack Sparrow				
Pitfall: The Lost Expedition				
Planetarian: Chiisana Hoshi no Yume	2006	Key	Prototype	JP
Playboy: The Mansion				
The Polar Express				
Pool Paradise				
Pool Paradise: International Edition				
Pool Shark 2				
Pop'n Music 7	2002	Konami	Konami	JP
Pop'n Music: Best Hits	2003	Konami	Konami	JP
Pop'n Music 8	2003	Konami	Konami	JP
Pop'n Music 9	2004	Konami	Konami	JP
Pop'n Music 10	2004	Konami	Konami	JP

Pop'n Music 11	2005	Konami	Konami	JP
Pop'n Music 12: Iroha	2006	Konami Digital Entertainment	Konami Digital Entertainment	JP
Pop'n Music 13: Carnival	2006	Konami Digital Entertainment	Konami Digital Entertainment	JP
Pop'n Music 14: Fever	2007	Konami Digital Entertainment	Konami Digital Entertainment	JP
Popeye: Hush Rush for the Spinach		Namco		
Portal Runner				
Power Rangers Dino Thunder				
Power Volleyball				
Powerdrome		Argonaut Games		
Powerpuff Girls: Relish Rampage				
Predator: Concrete Jungle				
Primal				
Prince of Persia: The Sands of Time		UbiSoft		
Prince of Persia: The Two Thrones		UbiSoft		
Prince of Persia: Warrior Within		UbiSoft		
Premier Manager 2005/2006				
Premier Manager 2006-2007				
Pride FC				
Pro Evolution Soccer	2001	Konami		
Pro Evolution Soccer 2	2002	Konami		
Pro Evolution Soccer 3	2003	Konami		
Pro Evolution Soccer 4	2004	Konami		
Pro Evolution Soccer 5	2005	Konami		
Pro Evolution Soccer 6	2006	Konami		
Pro Evolution Soccer 7	2007	Konami		
Pro Evolution Soccer Management	2005	Konami		
Pro Evolution Soccer 2008	2007	Konami		
Pro Evolution Soccer 2009	2008	Konami		
Pro Evolution Soccer 2010	2009	Konami		
Pro Evolution Soccer 2011	2010	Konami		
Pro Race Driver (TOCA Race Driver in Europe)		Codemasters		
Pro Rally 2002				
Project Altered Beast				
Project Eden				
Project Minerva				
Project: Snowblind				
ProStroke Golf- World Tour 2007				

Psi-Ops: The Mindgate Conspiracy				
Psychonauts	2005	Double Fine / Budcat	Majesco	NA
Pump It Up Exceed Special Edition				
The Punisher				
Putt Nutz		Black Mountain Games		
Puzzle Challenge: Crosswords and More				

Q

Title and source	Year	Developer(s)	Publisher(s)	Regions released
Q-Ball Billiards Master	2000	Ornith, ASK Corporation	Take-Two Interactive	
Quake III: Revolution	2001	Gremlin Interactive	Virgin Interactive	

R

Title and source	Year	Developer(s)	Publisher(s)	Regions released
Ratchet & Clank	2002	Insomniac Games	Sony Computer Entertainment	JP NA EU
Ratchet & Clank: Going Commando	2003	Insomniac Games	Sony Computer Entertainment	JP NA EU
Ratchet & Clank: Up Your Arsenal	2004	Insomniac Games	Sony Computer Entertainment	JP NA EU
Ratchet: Deadlocked	2005	Insomniac Games	Sony Computer Entertainment	JP NA EU AUS
Ratchet & Clank: Size Matters	2008	High Impact Games	Sony Computer Entertainment	NA EU
Ratchet and Clank Future: Quest for Booty	2009			
Rayman 2: Revolution	2000	Ubisoft	Ubisoft	
Rayman Arena/Rayman M	2001	Ubisoft	Ubisoft	
Rayman 3: Hoodlum Havoc	2003	Ubisoft	Ubisoft	
Rayman Raving Rabbids	2006	Ubisoft	Ubisoft	NA EU AUS
Raw Danger	2006	Irem	Agetec	JP NA EU AUS
Real World Golf				
Realm of the Dead				
Rebel Raiders: Operation Nighthawk	2006	Kando Games		
Red Card				
Red Dead Revolver	2004	Rockstar San Diego	Rockstar Games	JP NA EU
Red Faction	2001	Volition, Inc.	THQ	NA
Red Faction II	2002	Volition, Inc.	THQ	NA
Red Ninja: End of Honor	2005	Tranji Studios	Vivendi Universal Games	
The Red Star	2007	Acclaim Entertainment	XS Games	
Reel Fishing III	2003	Victor Interactive	Natsume	

Reign Of Fire	2002	Kuju Entertainment	BAM! Entertainment	
Reservoir Dogs	2004	Volatile Games	Eidos Interactive, Lionsgate	
Resident Evil 4	2005	Capcom	Capcom	JP NA EU AUS
Resident Evil Code: Veronica X	2001	Nex Entertainment	Capcom	
Resident Evil: Dead Aim	2003	Capcom	Capcom	
Resident Evil Outbreak	2003	Capcom	Capcom	
Resident Evil Outbreak: File#2	2004	Capcom	Capcom	
Resident Evil: Survivor 2 Code: Veronica	2001	Capcom	Capcom	
Return to Castle Wolfenstein: Operation Resurrection	2003	Gray Matter Interactive	Activision	
Rez	2001	United Game Artists	Sega, Sony Computer Entertainment	
Ribbit King	2003	Bandai	Bandai	
Richard Burns Rally	2004	Warthog Games	SCi	
Ricky Ponting Cricket 2007				AUS
Ridge Racer V	2000	Namco	Namco	
Ring of Red	2000	Konami	Konami	JP NA EU
Rise of the Kasai	2005	BottleRocket Entertainment	Sony Computer Entertainment	
Jet Li: Rise to Honor	2004	SCEA	SCEA	
Risk: Global Domination	2003	Cyberlore Studios	Atari	
River King: A Wonderful Journey	2005	Marvelous Interactive	Marvelous Interactive, Natsume	
Roadkill				
Road Trip Adventure	2002	E-game	Takara	
Robin Hood: Defender of the Crown	2003	Cinemaware	Capcom	NA EU
Robotech: Battlecry	2002	Vicious Cycle Software	TDK Mediactive	NA
Robotech Invasion	2004	Vicious Cycle Software	Global Star	NA
Robot Alchemical Drive	2002	Enix	Sandlot	JP NA
Robot Warlords				
Robot Wars: Arenas of Destruction	2001	Climax Entertainment	BBC Multimedia	
Rock Band	2007	Pi Studios	MTV Games, Electronic Arts	NA EU AUS
Rock Band 2	2008	Pi Studios	MTV Games	
Rocket Power: Beach Bandits	2002	THQ	THQ	
Rockman 2: The Power Fighters		Capcom		
Rocky Legends	2004	Venom Games	Ubisoft	
Rogue Galaxy	2005	Level-5	Sony Computer Entertainment	JP NA EU
Rogue Ops	2003	Bits Studios	Kemco	
Rogue Trooper	2006	Rebellion Developments	Eidos Interactive	
Roland Garros 2002				
Rollercoaster World				

Romance of the Three Kingdoms VII	2000	Koei	Koei	
Romance of the Three Kingdoms VIII	2003	Koei	Koei	JP NA EU
Romance of the Three Kingdoms IX	2003	Koei	Koei	
Romance of the Three Kingdoms X	2005	Koei	Koei	
Romance of the Three Kingdoms XI	2006	Koei	Koei	
Romancing SaGa: Minstrel Song	2005	Square Enix	Square Enix	
RPG Maker 2	2002	Kuusou Kagaku	Enterbrain, Agetec	
RPG Maker 3	2004	Run Time	Enterbrain, Agetec	
RPM Tuning	2004	Babylon Software	Wanadoo Edition	
R-Type Final	2003	Irem	Irem, Eidos Interactive	
Ruff Trigger: The Vanocore Conspiracy	2006	Playstos	Digital Publishing, Natsume	
EA Sports Rugby	2001	The Creative Assembly	Eletronic Arts	
Rugby 2004	2003	HB Studios	Eletronic Arts	
Rugby 2005	2005	HB Studios	Eletronic Arts	
Rugby 06	2006	HB Studios	Eletronic Arts	
Rugby 08	2007	HB Studios	Eletronic Arts	
Rugby League	2003	Sidhe Interactive	Home Entertainment Suppliers, Tru Blu Entertainment	EU AUS
Rugby League 2	2005	Sidhe Interactive	Home Entertainment Suppliers	EU AUS
Rugby League 2 World Cup Edition	2008	Sidhe Interactive	Home Entertainment Suppliers	EU AUS
Rugby League 3	2010	Sidhe Interactive	Home Entertainment Suppliers	EU AUS
Rule of Rose	2006	Punchline	Sony Computer Entertainment, Atlus	
Rumble Racing	2001	Eletronic Arts	Eletronic Arts	
Rumble Roses	2004	Konami	Konami	
Runabout 3				
Rune: Viking Warlord	2000	Human Head Studios	Gathering of Developers	
RLH: Run Like Hell	2002	Digital Mayhem	Capcom, Interplay Entertainment	
Rygar: The Legendary Adventure	2002	Tecmo	Tecmo	JP NA EU

S

Title and source	Year	Developer(s)	Publisher(s)	Regions released
S.L.A.I.: Steel Lancer Arena International	2005	Konami		JP NA
Sacred Blaze	2009	Flight-Plan	Banpresto	
Saint Seiya - Chapter Sanctuary				
Saint Seiya - The Hades				
Salt Lake 2002	2002	Attention To Detail	Eidos Interactive	JP NA EU AUS
Samurai Champloo: Sidetracked	2006	Bandai	Namco Bandai Games	JP NA

Samurai Jack: The Shadow of Aku	2004	Adrenium Games/Amaze Entertainment	Sega	NA EU
Samurai Showdown Anthology	2009	SNK Playmore	SNK Playmore	
Samurai Shodown V	2004	Yuki Enterprise	SNK Playmore	JP
Samurai Shodown VI	2006	SNK Playmore	SNK Playmore	JP
Samurai Warriors	2004	Omega Force	Koei	JP NA EU
Samurai Warriors: Xtreme Legends	2007	Omega Force	Koei	JP NA EU AUS
Samurai Warriors 2	2006	Omega Force	Koei	JP NA EU AUS
Samurai Warriors 2 Empires	2006	Omega Force	Koei	JP NA EU
Samurai Warriors 2: Xtreme Legends	2007	Omega Force	Koei	JP
Samurai Western	2005	Acquire	Spike/Atlus	JP NA EU
Saturday Night Speedway			Ratbag Games	
SBK-07 - Superbike World Championship	2007	Milestone	Black Bean Games	NA EU AUS
SBK Superbike World Championship	2009		Conspiracy Entertainment	
Scaler	2004	Artificial Mind and Movement	Global Star Software	NA EU
Scarface: The World is Yours	2006	Radical Entertainment	Sierra Entertainment	NA EU AUS
Scooby Doo: Mystery Mayhem	2004	Artificial Mind and Movement	THQ	NA EU
Scooby-Doo! Night of 100 Frights	2002	Heavy Iron Studios	THQ	NA EU
Scooby-Doo! Unmasked	2005	Artificial Mind and Movement	THQ	NA EU
Sea Monsters: A Prehistoric Journey	2008	DSI Games	National Geographic	
Second Sight	2004	Free Radical Design	Codemasters	
Secret Weapons Over Normandy	2003	Totally Games	LucasArts	
The Seed				
Seek and Destroy	2002	Takara	Conspiracy Entertainment	
Sega Ages			Sega	
Sega Classics Collection			Sega	
Sega Genesis Collection Sega Mega Drive Collection[JP EU]	2006	Digital Eclipse	Sega	
Sega Rally 2006	2006	Sega	Sega	
Sega Sports Tennis	2002	Hitmaker	Sega	
Sega Superstars	2004	Sonic Team	Sega	
Sensible Soccer 2006	2006			
Serious Sam: The Next Encounter				
Seven Samurai 20XX	2004	Dimps	Sega Sammy Studios	
Shadow Hearts	2001	Sacnoth	Aruze[JP], Midway Games	
Shadow Hearts: Covenant	2004	Nautilus	Aruze[JP], Midway Games	

Shadow Hearts: From The New World	2005	Nautilus	XSEED Games	JP NA EU
Shadow Man: 2econd Coming	2002	Acclaim Entertainment	Acclaim Entertainment	
Shadow of Destiny *Shadow of Memories*[JP EU]	2001	Konami	Konami	
Shadow of the Colossus	2005	Team Ico	SCEI	JP NA EU AUS
Shadow of Rome	2005	Capcom	Capcom	
Shadow the Hedgehog	2005	Sonic Team USA	Sega	
Shaman King: Funbari Spirits		Dimps	Bandai	
Shaman King: Power of Spirit		Konami		
SharkPort 2				
Shark Tale				
Shaun Palmer's Pro Snowboarder		Activision	Activision 02	
Shellshock: Nam '67		Guerrilla Games	Eidos Interactive	
Shifters				
Shinobi		Overworks	Sega	
Shinobido: Way of the Ninja		Acquire	Spike[JP], Sony Computer Entertainment [EU AU]	
Shin Megami Tensei: Devil Summoner: Raidou Kuzunoha vs. The Soulless Army		Atlus	Atlus, Koei[EU]	
Shin Megami Tensei: Digital Devil Saga		Atlus	Atlus	
Shin Megami Tensei: Digital Devil Saga 2		Atlus	Atlus	
Shin Megami Tensei: Nocturne		Atlus	Atlus	
Shin Megami Tensei: Persona 3	2006	Atlus	Atlus	JP NA
Shin Megami Tensei: Persona 4	2008	Atlus	Atlus	JP NA
Shining Dragon				
Shining Force Neo		Neverland, Amusement Vision	Sega	
Shining Tears		Nextech, Amusement Vision	Sega	
Showdown: Legends of Wrestling				
Shrek: Super Party				
Shrek Super Slam				
Shrek 2				
Shuffle! On the Stage				JP
Silent Hill 2			Konami	
Silent Hill 3			Konami	
Silent Hill 4: The Room			Konami	
Silent Hill: Origins	2008		Konami	NA EU
Silent Line: Armored Core	2005	From Software	Agetech	JP NA EU
Silpheed: The Lost Planet				

Silent Scope	2000	Konami		
Simple 2000 Series Vol. 01: The Table Game		Yuki Enterprise	D3 Publisher	JP
Simple 2000 Series Vol. 02: The Party Game		HuneX	D3 Publisher	JP EU
Simple 2000 Series Vol. 03: The Bass Fishing		Vingt-et-un Systems	D3 Publisher	JP NA EU
Simple 2000 Series Vol. 04: The Double Mahjong Puzzle		Metro Corporation	D3 Publisher	JP
Simple 2000 Series Vol. 05: The Block Kuzushi Hyper			D3 Publisher	JP EU
Simple 2000 Series Vol. 06: The Snowboard			D3 Publisher	JP EU
Simple 2000 Series Vol. 07: The Boxing - Real Fist Fight			D3 Publisher	JP EU
Simple 2000 Series Vol. 08: The Tennis		HuneX	D3 Publisher	JP EU
Simple 2000 Series Vol. 09: The Ren'ai Adventure - Bittersweet Fools		HuneX	D3 Publisher	JP
Simple 2000 Series Vol. 10: The Table Game Sekaihen		Yuki Enterprise	D3 Publisher	JP EU
Simple 2000 Series Vol. 42: The Ishu Kakutougi - Boxing vs Kick vs Karate vs Pro Wres vs Jujutsu vs...		Daft	D3 Publisher	JP EU
Simple 2000 Series Vol. 50: The Daibijin		Tamsoft	D3 Publisher	JP EU
Simple 2000 Series Vol. 61: The Oneechanbara		Tamsoft	D3 Publisher	JP EU
Simple 2000 Series Vol. 74: Onna no Ko Senyo - The Ouji-sama to Romance: Ripple no Tamago		HuneX	D3 Publisher	JP
Simple 2000 Series Vol. 80: The Oneechamploo		Tamsoft	D3 Publisher	JP EU
Simple 2000 Series Vol. 86: The Menko Shutoku Simulation - Dorokotsuho Taioban		Vingt-et-un Systems	D3 Publisher	JP
Simple 2000 Series Vol. 90: The Oneechanbara 2		Tamsoft	D3 Publisher	JP
Simple 2000 Series Vol. 105: The Maid Fuku to Kikanju		Rideon	D3 Publisher	JP
Simple 2000 Series Vol. 114: The Jo'okappichi Torimonocho - Oharu-chan Go Go Go!		Tamsoft	D3 Publisher	JP
The Simpsons Game	2007	EA Redwood Shores	Electronic Arts	JP NA EU AUS
The Simpsons Hit & Run	2003	Free Radical Design	Vivendi Games	
The Simpsons Road Rage	2001	Radical Entertainment	Vivendi Games	
The Simpsons Skateboarding	2002	The Code Monkeys	Electronic Arts	
The Sims	2002	Edge of Reality	Electronic Arts	NA
The Sims: Bustin' Out	2003	Maxis	Electronic Arts	
The Sims 2	2004	Maxis	Electronic Arts	
The Sims 2 Pets	2006	Maxis	Electronic Arts	
The Sims 2 Castaway	2007	Maxis	Electronic Arts	
SingStar	2004	London Studio	Sony Computer Entertainment	EU AUS
SingStar '80s	2005	London Studio	Sony Computer Entertainment	NA EU AUS
SingStar Anthems	2006	London Studio	Sony Computer Entertainment	EU AUS
SingStar Legends	2006	London Studio	Sony Computer Entertainment	NA EU AUS
SingStar Party	2004	London Studio	Sony Computer Entertainment	EU AUS
SingStar Pop	2005	London Studio	Sony Computer Entertainment	NA EU AUS

SingStar Rocks!	2006	London Studio	Sony Computer Entertainment	NA EU AUS
SingStar R&B	2007	London Studio	Sony Computer Entertainment	EU AUS
SingStar '90s	2007	London Studio	Sony Computer Entertainment	NA EU AUS
SingStar The Dome	2005	London Studio	Sony Computer Entertainment	NA EU AUS
SingStar Rock Ballads	2007	London Studio	Sony Computer Entertainment	EU AUS
Siren	2003	SCE Japan Studio	Sony Computer Entertainment	JP EU NA
Siren 2	2006	SCE Japan Studio	Sony Computer Entertainment	JP EU
Ski & Shoot	2009		Conspiracy Entertainment	
SkyGunner	2002	PixelArts	Atlus	
Sky Odyssey	2000	Cross	Activision	
S.L.A.I.: Steel Lancer Arena International				
Sled Storm				
Sly Cooper and the Thievius Raccoonus	2002	Sucker Punch Productions	Sony Computer Entertainment	
Sly 2: Band of Thieves	2004	Sucker Punch Productions	Sony Computer Entertainment	
Sly 3: Honor Among Thieves	2005	Sucker Punch Productions	Sony Computer Entertainment	JP NA EU AUS
Smash Court Tennis Pro Tournament				
Smash Court Tennis Pro Tournament 2	2004	Namco	Namco	
Smackdown vs. Raw				
Smarties: Meltdown	2006	Europress	Europress	EU
Smuggler's Run	2000	Angel Studios	Rockstar Games	
Smuggler's Run 2: Hostile Territory	2001	Rockstar Games	Rockstar Games	
The Sniper 2				
Sniper Elite				
SNK Arcade Classics Vol. 1			SNK Playmore	
Snoopy vs. The Red Baron				
Snowboard Racer 2				
SOCOM: U.S. Navy SEALs	2002	Zipper Interactive	SCEI	NA EU
SOCOM II: U.S. Navy SEALs	2003	Zipper Interactive	SCEI	NA EU
SOCOM 3: U.S. Navy SEALs	2005	Zipper Interactive	SCEI	NA EU
SOCOM U.S. Navy SEALs: Combined Assault	2006	Zipper Interactive	SCEI	NA
Soccer Life!				
Sol Divide				
Sonic Gems Collection	2005	Sonic Team	Sega	
Sonic Heroes	2004	Sonic Team	Sega	
Sonic Mega Collection Plus	2004	Sonic Team	Sega	
Sonic Riders	2006	Sonic Team	Sega	
Sonic Riders: Zero Gravity	2008	Sonic Team	Sega	
Sonic Unleashed	2008	Sonic Team	Sega	
The Sopranos: Road to Respect				

Soul Calibur II	2003	Namco	Namco	JP NA EU
Soul Calibur III	2005	Namco	Namco	
Space Channel 5: Special Edition			Sega	
Space Invaders: Invasion Day	2002	Taito Corporation	Taito Corporation	
Spartan: Total Warrior	2005	The Creative Assembly	Sega	
Spawn: Armageddon				
Sphinx and the Cursed Mummy	2003	Eurocom	THQ	
The Spiderwick Chronicles	2008	Stormfront Studios	Sierra Entertainment	
Spider-Man	2002	Treyarch	Activision	
Spider-Man 2	2004	Treyarch	Activision	
Spider-Man 3	2007	Vicarious Visions	Activision	
Spider-Man: Web of Shadows - Amazing Allies Edition	2008		Activision	
Splashdown	2001	Rainbow Studios	Atari	
SpongeBob SquarePants: Battle for Bikini Bottom	2001		THQ	
SpongeBob SquarePants: Creature from the Krusty Krab	2006		THQ	
SpongeBob SquarePants: Revenge of the Flying Dutchman	2002		THQ	
The SpongeBob SquarePants Movie (video game)	2004		THQ	
SpongeBob SquarePants: Lights, Camera, Pants!	2005		THQ	
Sprint Cars: Road to Knoxville				
Spy Fiction				
Spy Hunter				
Spy Hunter: Nowhere to Run				
Spy Hunter 2				
Spyro: A Hero's Tail				
Spyro: Enter the Dragonfly				
SSX			Electronic Arts	
SSX 3			Electronic Arts	
SSX on Tour			Electronic Arts	
SSX Tricky			Electronic Arts	
Stacked with Daniel Negreanu				
Stargate SG-1: The Alliance				
Star Ocean: Till the End of Time	2003	Tri-Ace	Square Enix	JP NA EU
Star Trek: Conquest	2007	4J Studios	Bethesda Softworks	NA EU
Star Trek: Encounters	2006	4J Studios	Bethesda Softworks	NA
Star Trek: Shattered Universe	2004	Starsphere Interactive	TDK Mediactive	NA
Star Trek: Voyager Elite Force				
Star Wars: Battlefront			LucasArts	
Star Wars: Battlefront II	2005	Pandemic	LucasArts	NA EU
Star Wars: Bounty Hunter			LucasArts	

Star Wars: The Clone Wars			LucasArts	
Star Wars Episode III: Revenge of the Sith			LucasArts	
Star Wars: Jedi Starfighter			LucasArts	
Star Wars: Starfighter			LucasArts	
Star Wars: The Force Unleashed			LucasArts	
Starsky and Hutch				
State of Emergency	2002	VIS Entertainment Ltd.	Rockstar Games	NA EU
State of Emergency 2	2006	DC Studios	SouthPeak Interactive	NA
Stealth Force: The War on Terror				
Steambot Chronicles				
Steamboy			Bandai	
Steel Dragon EX				
Stella Deus: The Gate of Eternity				
Stock Car Crash				
Stolen				
The Street Basketball: 3 on 3	2003	Tamsoft	D3 Publisher	JP EU
Street Racing				
Street Fighter Anniversary Collection			Capcom	
Street Fighter Alpha Anthology				
Street Fighter EX3			Capcom	
Stretch Panic				
Strike Force Bowling				
Stuntman			Atari	
Stuntman Ignition		THQ		
Sub Rebellion				
The Suffering				
The Suffering: Ties That Bind				
Suggoi! Arcana Heart 2	2009	Examu	AQ Interactive	JP
Suikoden III			Konami	
Suikoden IV			Konami	
Suikoden V			Konami	
Suikoden Tactics			Konami	
The Sum of All Fears				
Summer Heat Beach Volleyball				
Summoner				
Summoner 2				
Super Dragon Ball Z				
Super Robot Wars Alpha 2				JP
Super Robot Wars Alpha 3				JP

Super Robot Wars Impact				JP
Super Robot Wars MX				JP
Supercar Street Challenge				
SuperLite 2000: Tokyo Bus Guide				JP
Superman Returns	2006	EA Tiburon	Electronic Arts	
Super Monkey Ball Adventure	2006	Traveller's Tales	Sega	
Super Monkey Ball Deluxe	2005	Sega	Sega	
Super Trucks Racing	2002	Jester Interactive	Jester Interactive/XS Games	
SVC Chaos: SNK vs. Capcom	2003	SNK Playmore	SNK Playmore	
Sven-Göran Eriksson's World Challenge	2002	3DO	3DO	EU
SWAT: Global Strike Team	2003	Argonaut Games	Argonaut Games/Sierra Entertainment	
The Sword of Etheria	2005	KCET	Konami	
Syphon Filter: The Omega Strain	2004	Sony Bend	Sony Computer Entertainment	
Syphon Filter: Dark Mirror	2007	Sony Bend	Sony Computer Entertainment	

T

Title and source	Year	Developer(s)	Publisher(s)	Regions released
Taito Legends	2005	Taito Corporation	Sega	JP NA EU
Taito Legends 2	2006	Taito Corporation	Destineer, Empire Interactive	
Tak and the Power of Juju	2003	Avalanche Software	THQ	
Tak 2: The Staff of Dreams	2004	Avalanche Software	THQ	
Tak: The Great Juju Challenge	2005	Avalanche Software	THQ	
Tak and the Guardians of Gross	2008	THQ	THQ	
Tales of the Abyss	2005	Namco Tales Studio	Namco	JP NA
Tales of Destiny (Remake)	2006	Wolfteam	Namco	JP
Tales of Destiny 2	2002	Telenet Japan, Wolfteam	Namco	JP NA
Tales of Legendia	2005	Namco	Namco	JP NA
Tales of Rebirth	2004	Namco Tales Studio	Namco	JP
Tales of Symphonia	2004	Namco Tales Studio	Namco	JP NA
Taz Wanted	2002	Blitz Games	Infogrames	
Technic Beat	2002	Arika	Mastiff	
Tekken Tag Tournament	2000	Namco	Namco	
Tekken 4	2002	Namco	Namco	JP NA EU
Tekken 5	2005	Namco	Namco	JP NA EU AUS
Teen Titans	2006	Artificial Mind and Movement	THQ	
Teenage Mutant Ninja Turtles	2003	Konami	Konami	
Teenage Mutant Ninja Turtles 2: Battle Nexus	2004	Konami	Konami	

Teenage Mutant Ninja Turtles 3: Mutant Nightmare	2005	Konami	Konami	
Tenchu: Fatal Shadows	2005	K2 LLC	Sega	JP NA EU AUS
Tenchu: Wrath of Heaven	2003	K2 LLC	Activision	JP NA EU
Tenshi no Present	2000	Nippon Ichi	Nippon Ichi	JP
Terminator 3: Rise of the Machines	2003	Black Ops Entertainment	Atari	
Terminator 3: The Redemption	2004	Paradigm Entertainment	Atari	
Test Drive	2001	Pitbull Syndicate	Atari	
Test Drive: Eve of Destruction	2004	Monster Games	Atari	
Test Drive Off-Road: Wide Open	2001	Angel Studios	Infogrames	
Test Drive Unlimited	2007	Melbourne House	Atari	NA EU AUS
Tetris Worlds	2002	Blue Planet Software	THQ	NA
The Super Dimension Fortress Macross	2003	Sega-AM2	Bandai	JP
The Arcade	2005	Liquid Games		
The Thing	2002	Computer Artworks	VU Games	
Theme Park World	2000	Bullfrog Productions	EA	
This Is Football 2002	2002	SCE London Studio	SCEE	EU
This Is Football 2003	2003	SCE London Studio	SCEE	EU
This Is Football 2004	2004	SCE London Studio	SCEE	EU
This Is Football 2005	2005	SCE London Studio	SCEE	EU
Thomas & Friends: A Day at the Races	2007	Broadsword Interactive	Blast! Entertainment	
Thrillville	2006	Frontier Developments	LucasArts	NA EU
Thunderstrike: Operation Phoenix	2001	Core Design	Eidos Interactive	
Thunder Force VI	2008		Sega	JP
Tiger Woods PGA Tour 2001	2000	Electronic Arts	EA Sports	NA
Tiger Woods PGA Tour 2002	2001	Electronic Arts	EA Sports	NA
Tiger Woods PGA Tour 2003	2002	Electronic Arts	EA Sports	NA
Tiger Woods PGA Tour 2004	2003	Electronic Arts	EA Sports	NA
Tiger Woods PGA Tour 2005	2004	Electronic Arts	EA Sports	NA
Tiger Woods PGA Tour 2006	2005	Electronic Arts	EA Sports	NA
Tiger Woods PGA Tour 07	2006	Electronic Arts	EA Sports	JP NA EU AUS
Tiger Woods PGA Tour 08	2007	Electronic Arts	EA Sports	JP NA EU AUS
Tiger Woods PGA Tour 09	2008	Electronic Arts	EA Sports	JP NA EU AUS
Tiger Woods PGA Tour 10	2009	Electronic Arts	EA Sports	
Tiger Woods PGA Tour 11	2010	Electronic Arts	EA Sports	JP NA EU AUS
Time Crisis II	2001	Namco	Namco	JP NA EU
Time Crisis 3	2003	Nextech	Namco	JP NA EU
Time Crisis: Crisis Zone	2004	Namco	Namco	JP NA
Time Traveler	2001	Sega	Digital Leisure	NA
TimeSplitters	2000	Free Radical Design	Eidos Interactive	NA

TimeSplitters 2	2002	Free Radical Design	Eidos Interactive	JP NA EU
TimeSplitters: Future Perfect	2005	Free Radical Design	EA Games	NA EU
Tiny Toon Adventures: Defenders of the Looniverse		Treasure	Conspiracy Games	
TMNT	2007	Ubisoft Montreal	UbiSoft	NA EU AUS
TOCA Race Driver	2002	Codemasters	Codemasters	
TOCA Race Driver 2	2004	Codemasters	Codemasters	
TOCA Race Driver 3	2006	Codemasters	Codemasters	NA EU
ToHeart2	2004		Aquaplus	JP
Tokimeki Memorial 3	2001	Konami	Konami	JP
Tokimeki Memorial Girl's Side	2002	Konami	Konami	JP
Tokobot Plus: Mysteries of the Karakuri	2006	Tecmo	Tecmo	
Tokyo Road Race				
Tokyo Road Racer				
Tokyo Xtreme Racer: 3	2004	Genki	Crave Entertainment	JP NA
Tokyo Xtreme Racer: Drift	2003	Genki	Crave Entertainment	JP NA
Tokyo Xtreme Racer: Drift 2	2005	Genki	Genki, Crave Entertainment	JP NA
Tokyo Xtreme Racer: Zero	2001	Genki	Crave Entertainment	JP NA EU AUS
Tomb Raider: The Angel of Darkness	2003	Core Design	Eidos Interactive	JP NA EU AUS
Tomb Raider: Legend	2006	Crystal Dynamics	Eidos Interactive	JP NA EU AUS
Tomb Raider: Anniversary	2007	Crystal Dynamics	Eidos Interactive	JP NA EU AUS
Tomb Raider: Underworld	2008	Crystal Dynamics	Eidos Interactive	JP NA EU AUS
Tom Clancy's Ghost Recon	2002	Red Storm Entertainment	UbiSoft	
Tom Clancy's Ghost Recon: Advanced Warfighter	2006	UbiSoft, Red Storm Entertainment	UbiSoft	JP NA EU AUS
Tom Clancy's Ghost Recon: Jungle Storm	2004	Red Storm Entertainment	UbiSoft	
Tom Clancy's Ghost Recon 2	2004	Red Storm Entertainment	UbiSoft	
Tom Clancy's Rainbow Six 3: Raven Shield	2004	UbiSoft, Red Storm Entertainment	UbiSoft	
Tom Clancy's Splinter Cell	2003	UbiSoft	UbiSoft	
Tom Clancy's Splinter Cell: Chaos Theory	2005	UbiSoft	UbiSoft	
Tom Clancy's Splinter Cell: Double Agent	2006	UbiSoft	UbiSoft	JP NA EU AUS
Tom Clancy's Splinter Cell: Pandora Tomorrow	2004	UbiSoft	UbiSoft	
Tomoyo After: It's a Wonderful Life CS Edition	2007	Key	Prototype	JP
Tony Hawk's Pro Skater 3	2001	Neversoft	Activision	
Tony Hawk's Pro Skater 4	2002	Neversoft	Activision	
Tony Hawk's Underground	2003	Neversoft	Activision	
Tony Hawk's Underground 2	2004	Neversoft	Activision	
Tony Hawk's American Wasteland	2005	Neversoft	Activision	
Tony Hawk's Project 8	2006	Shaba Games	Activision	NA EU AUS

Tony Hawk's Downhill Jam	2007	SuperVillain Studios	Activision	NA EU AUS
Tony Hawk's Proving Ground	2007	Page 44 Studios	Activision	JP NA EU AUS
Torino 2006	2006	49Games	2K Sports	
Total Overdose: A Gunslinger's Tale in Mexico	2005	Deadline Games	Sci Entertainment, Eidos Interactive	
Top Spin	2005	PAM Development	2K Sports	
Tourist Trophy	2006	Polyphony Digital	Sony Computer Entertainment	JP NA EU AUS
Train Simulator Real	2001	Sony Computer Entertainment	Sony Computer Entertainment	JP
Transformers	2004	Melbourne House	Atari	NA EU
Transformers: The Game	2007	Traveller's Tales	Atari	NA EU
Trapt	2005	Tecmo	Tecmo	
Treasure Planet	2004	Bizarre Creations	SCEA	
Tribes: Aerial Assault	2002	Inevitable	Sierra Entertainment	NA EU
Trigger Man	2004	Point of View	Crave Entertainment	
Trivial Pursuit Unhinged	2004	Artech Studios	Atari	
Trivial Pursuit	2009	Electronic Arts	Electronic Arts	
Truck Racing 2	2005	Midas Interactive		
True Crime: New York City	2005	Luxoflux	Activision	
True Crime: Streets of LA	2003	Luxoflux	Activision	
Tsugunai: Atonement	2001	Cattle Call	Atlus	JP NA
Turok: Evolution	2002	Acclaim Entertainment	Acclaim Entertainment	
Twinkle Star Sprites: La Petite Princesse	2004	SNK Playmore	SNK Playmore	JP
Twisted Metal: Black	2001	Incog Inc.	Sony Computer Entertainment	
Twisted Metal: Black Online	2002	Incog Inc.	Sony Computer Entertainment	
Ty the Tasmanian Tiger	2002	Krome Studios	Electronic Arts	
Ty the Tasmanian Tiger 2: Bush Rescue	2004	Krome Studios	Electronic Arts	
Ty the Tasmanian Tiger 3: Night of the Quinkan	2005	Krome Studios	Activision	

U

Title	Year	Developer(s)	Publisher(s)	Regions released
UEFA Champions League 2004-2005	2004	EA Sports	EA	
UEFA Euro 2004				
UFC Sudden Impact	2001	Genki	Crave Entertainment	JP NA EU
UFC: Throwdown				
Ultimate Alliance	2006	Raven Software	Activision	NA EU AUS
Ultimate Board Game Collection				
Ultimate Mind Games				
Ultimate Sky Surfer				
Ultimate Spider-Man	2005	Treyarch	Activision	NA EU

Ultraman				
Ultraman Fighting Evolution 2				
Ultraman Fighting Evolution 3				
Ultraman Fighting Evolution Rebirth				
Ultraman Nexus	2005	Bandai, RenderWare	Bandai	
Under the Skin	2004	Capcom	Capcom	JP NA EU
Underworld: The Eternal War	2004	Lucky Chicken Games Inc.	Play it! Limited	EU
UNiSON	2000	Tecmo	Tecmo	JP NA
Unlimited SaGa	2002	Square Co.	Square Co., Square Enix, Atari Europe	JP NA EU
Unreal Tournament	2000	EpicGames / DigitalExtremes	Infogrames	NA EU
Urban Chaos: Riot Response	2006	Rocksteady Studios	Eidos Interactive	NA EU
Urban Reign	2005	Namco	Namco	JP NA EU
The Urbz: Sims in the City	2004	Maxis	Electronic Arts	JP NA EU

V

Title	Year	Developer(s)	Publisher(s)	Regions released
V8 Supercar: Race Driver	2003	Atari	Codemasters	
V8 Supercar 2	2005	Atari	Codemasters	
V8 Supercar 3	2006	Atari	Codemasters	
Valkyrie Profile 2: Silmeria	2006	tri-Ace	Square Enix	JP NA EU
Vampire: Darkstalkers Collection	2005	Capcom	Capcom	JP
Vampire Night	2001	AM1	Namco	
Vampire Panic				
Van Helsing	2004	Saffire	VU Games	
VeggieTales: Larry-Boy and the Bad Apple		Papaya Studio	Crave Entertainment	
Venus And Braves				
Vexx	2003	Acclaim Entertainment	Acclaim Entertainment	
Vib-Ripple	2004	NanaOn-Sha	Sony Computer Entertainment	
Victorious Boxers: Ippo's Road To Glory	2000	New Corporation	Entertainment Software Publishing, Empire Interactive	JP NA EU
Victorious Boxers 2	2004	New Corporation	Entertainment Software Publishing, Empire Interactive, VU Games	JP NA EU
Vietcong Purple Haze	2004	Coyote	Gathering of Developers	
Viewtiful Joe	2004	Clover Studio	Capcom	
Viewtiful Joe 2	2004	Clover Studio	Capcom	
V.I.P.				

Virtua Cop: Elite Edition	2002	Sega-AM2	Acclaim	
Virtua Fighter 4	2002	Sega-AM2	Sega	JP NA EU
Virtua Fighter 4 Evolution	2003	Sega-AM2	Sega	JP NA EU AUS
Virtua Quest	2004	Sega-AM2	Sega	JP NA
Virtua Tennis 2	2002	Hitmaker	Sega	JP NA EU
Virtual-On Marz	2003	Hitmaker		JP NA
Visual Mix Ayumi Hamasaki Dome Tour 2001				
V-Rally 3	2002	Eden Studios	Atari	

W

Title	Year	Developer(s)	Publisher(s)	Regions released
Wacky Races Starring Dastardly and Muttley	2001	Infogrames	Infogrames	
Wallace & Gromit in Project Zoo	2003	Frontier Developments	BAM! Entertainment	
Wallace & Gromit: The Curse of the Were-Rabbit	2005	Frontier Developments	Konami	
War of the Monsters	2003	Incognito Entertainment	Sony Computer Entertainment	NA EU JP
The Warriors	2005	Rockstar Toronto	Rockstar Games	NA EU
Warriors Orochi	2007	Omega Force	Koei	JP NA EU AUS
Warriors Orochi 2	2008	Omega Force	Koei	JP NA EU
Way of the Samurai	2002	Acquire	Spike, BAM! Entertainment, Eidos Interactive	JP NA EU
Way of the Samurai 2	2003	Acquire	Spike, Capcom	JP NA EU
Warhammer 40,000: Fire Warrior	2003	Kuju Entertainment	THQ	
Warship Gunner 2				
We Love Katamari	2005	Namco	Namco, Electronic Arts	JP NA EU AUS KOR
Whiplash	2003	Crystal Dynamics	Eidos Interactive	NA EU AUS
Who Wants To Be A Millionaire 2nd Edition				
Wild Arms 3	2002	Media Vision	Sony Computer Entertainment	JP NA EU
Wild Arms 4	2005	Media Vision	SCEI, XSEED Games, 505 Games	JP NA EU AUS
Wild Arms 5	2006	Media Vision	Sony Computer Entertainment, XSEED Games, 505 Games	JP NA EU
Wild Arms Alter Code: F	2003	Media Vision	SCEI, Agetec	JP NA
Wild Wild Racing	2000	Rage Software Limited	Interplay, Imagineer	JP NA EU
Winter Sports 2: The Next Challenge	2008	49 Games	Conspiracy Entertainment	NA EU
WinBack: Covert Operations	2001	Omega Force	Koei	
WinBack 2: Project Poseidon	2006	Cavia, inc.	Koei	
Wipeout Fusion	2002	Sony Studio Liverpool	SCEE Bam! Entertainment	EU NA

Without Warning	2005	Circle Studio	Capcom	
Wizardry: Tale of the Forsaken Land	2001	Racjin	Atlus	JP NA EU
Woody Woodpecker	2001			
World Championship Poker	2005	Coresoft	Crave Entertainment	NA
World Championship Poker: Featuring Howard Lederer - All In	2006	Coresoft	Crave Entertainment	NA
World Championship Snooker 2001				
World Destruction League: Thunder Tanks	2000	3DO	3DO	
World of Outlaws: Sprint Cars 2002		Ratbag Games	Infogrames	
World Poker Tour	2006	2K Sports	2K Sports	
World Racing 2	2005	Synetic		
World Rally Championship	2001	Evolution Studios	Sony Computer Entertainment	NA
World Rally Championship 2	2002	Evolution Studios	Sony Computer Entertainment	NA
World Rally Championship 3	2003	Evolution Studios	Sony Computer Entertainment	JP EU
World Rally Championship 4	2004	Evolution Studios	Sony Computer Entertainment	JP EU
World Series of Poker	2005	Left Field Productions	Activision	NA
World Series of Poker: Tournament of Champions	2006	Left Field Productions	Activision	NA
World Soccer Winning Eleven 9 Pro Evolution Soccer 5 *PAL*	2005	Konami	Konami	JP EU NA
World Super Police	2006	Midas Interactive Entertainment	Jaleco	
World Tour Soccer 2005				
World War I: Aces Of the Sky	2006	Midas Interactive Entertainment	NAPS Team	
World War II: Battle over the Pacific	2006	Midas Interactive Entertainment	Midas Interactive Entertainment	
World War II:Soldier	2006	Atomic Planet Entertainment	Midas Interactive Entertainment	
World War II: Tank Battles	2006		Midas Interactive Entertainment	
World War Zero	2004	4X Studios Kylotonn	Dreamcatcher Interactive Wanadoo Edition	
Worms 3D	2003	Team17	Acclaim Sega	EU NA
Worms 4: Mayhem	2005	Team17	Majesco Codemasters	EU
Worms Forts: Under Siege	2004	Team17	Sega	EU NA
Wrath Unleashed	2004	The Collective, Inc.	LucasArts	EU NA
WRC: Rally Evolved	2005	Evolution Studios	SCE	EU NA
WWE Crush Hour	2003	Pacific Coast Power and Light	THQ	EU NA
WWE SmackDown! Here Comes The Pain	2003	Yukes	THQ Yukes	JP EU NA

WWF SmackDown! Just Bring It	2001	Yukes	THQ Yukes	JP EU NA
WWE SmackDown! Shut Your Mouth	2002	Yukes	THQ Yukes	NA
WWE SmackDown! vs. RAW	2004	Yukes	THQ Yukes	EU NA
WWE SmackDown! vs. RAW 2006	2005	Yukes	THQ Yukes	JP EU NA
WWE SmackDown vs. Raw 2007	2006	Yukes	THQ	JP EU NA AUS
WWE SmackDown vs. Raw 2008	2007	Yukes	THQ	JP EU NA AUS
WWE Smackdown vs. Raw 2009	2008	Yukes	THQ	EU NA
WWE SmackDown vs. Raw 2010	2009	Yukes	THQ	JP EU NA AUS

Title	Year	Developer(s)	Publisher(s)	Regions released
X2: Wolverine's Revenge	2003	Raven Software	Activision	NA EU
The X-Files: Resist or Serve	2004	Black Ops Entertainment	Vivendi Universal	NA EU
X-Men Legends	2004	Raven Software	Activision	NA
X-Men Legends II: Rise of Apocalypse	2005	Raven Software	Activision	NA
X-Men: Next Dimension	2002	Paradox Development	ActiVision	NA
X-Men Origins: Wolverine	2009	Raven Software, Amaze Entertainment	ActiVision	
X-Squad	2000	EA	EA	NA EU
X-Men: The Official Game	2006	Z-Axis	Activision	NA
Xenosaga Episode I: Der Wille zur Macht	2003	Monolith Soft	Namco	JP NA
Xenosaga Episode II: Jenseits von Gut und Böse	2005	Monolith Soft	Namco	JP NA EU
Xenosaga Episode III: Also sprach Zarathustra	2006	Monolith Soft	Namco Bandai Games	JP NA
XG3: Extreme G Racing	2001	Acclaim Entertainment	Acclaim Entertainment	NA
XGRA: Extreme-G Racing Association	2003	Acclaim Cheltenham	Acclaim Entertainment	NA
Xiaolin Showdown	2006	BottleRocket Entertainment	Konami	NA
Xtreme Express	2004	Syscom Entertainment	Midas Interactive Entertainment	JP EU
XIII	2003	Ubisoft Paris	Ubisoft	NA EU

Y

Title and source	Year	Developer(s)	Publisher(s)	Regions released
Yakuza	2006	Sega	Sega	JP NA EU
Yakuza 2	2008	Sega	Sega	JP NA EU
Yanya Caballista: City Skater	2001	Koei	Koei	NA
Yourself!Fitness	2005	ResponDESIGN	ResponDESIGN	NA
Ys I And II Eternal Story	2003	Nihon Falcom	Nihon Falcom	JP
Ys III: Wanderers from Ys	2005	Nihon Falcom	Nihon Falcom	JP
Ys IV: Mask of the Sun	2005	Nihon Falcom	Nihon Falcom	JP
Ys V: Kefin, The Lost City of Sand	2006	Nihon Falcom	Nihon Falcom	JP
Ys: The Ark of Napishtim	2005	Nihon Falcom	Konami	JP NA EU
Yu-Gi-Oh! Capsule Monster Coliseum	2005	Konami	Konami	JP NA
Yu-Gi-Oh! The Duelists of the Roses	2003	Konami	Konami	JP NA
Yu-Gi-Oh GX! Tag Force Evolution	2008	Konami	Konami	JP NA
Yu Yu Hakusho: Dark Tournament	2004	Digital Fiction	Atari	NA
Yu Yu Hakusho Forever	2005	Konami	Konami	JP

Z

Title	Year	Developer(s)	Publisher(s)	Regions released
Zapper: One Wicked Cricket	2002	Blitz Games	Infogrames	NA EU
Zatch Bell! Mamodo Battles	2005	8ing	Bandai	JP NA
Zatch Bell! Mamodo Fury	2006	Mechanic Arms	Namco Bandai Games	JP NA
Zoids Infinity Fuzors	2005	Tomy	Tomy	JP
Zoids Struggle	2004	Tomy	Tomy	JP
Zoids Tactics	2005	Tomy	Tomy	JP
Zombie Hunters	2006	Tamsoft	Essential Games	JP EU
Zombie Virus	2006	Vingt-et-un Systems	Essential Games	JP EU
Zombie Zone	2005	Tamsoft	505 Game Street	JP EU
Zone of the Enders	2001	KCEJ	Konami	JP NA EU
Zone of the Enders: The 2nd Runner	2003	Konami	Konami	JP NA EU

See also

- Chronology of PlayStation 2 games
- → List of PlayStation 2 games with HD support
- List of Xbox games

References

[1] " PlayStation 2 Game List - Category: Action (http://www.gamefaqs.com/console/ps2/cat_54.html)". GameFAQs. . Retrieved 2009-03-02.

List of PlayStation 2 DVD-9 games

This is a list of Sony → PlayStation 2 games that use the DVD-9 (dual layer) format.

Contents

0-9

- 24: The Game ( )
- 50 Cent: Bulletproof ( )

A

- ATV Offroad Fury 4 ( )

C

- Champions of Norrath ( )

D

- Dynasty Warriors 6 (( )

F

- Fight Night Round 3 (( )
- Forbidden Siren 2 ( )

G

- Genji: Dawn of the Samurai ( )
- God of War ( )
- God of War II ( )
- Gran Turismo 4 ( )
- Gran Turismo Concept: 2002 Tokyo-Geneva ( )
- Guitar Hero: Metallica ( )
- Guitar Hero World Tour ( )
- The Guy Game ( )

K

- Kessen III ( )

M

- Madden NFL 2005 Collector's Edition ( )
- Medal of Honor: Rising Sun ( )
- Metal Gear Solid 2: Substance ( )
- Midnight Club 3: DUB Edition Remix ( )

R

- Rock Band ( )
- Rock Band 2 ( )
- Rogue Galaxy ( )
- Ryu Ga Gotoku 2 (Playstation The Best) ( )

S

- Sakura Taisen 5: Saraba, Itoshiki Hito yo ( )
- Sakura Taisen: Atsuki Chishio Ni ( )
- Samurai Warriors 2 Xtreme Legends ( )
- School Days LxH ( )
- SingStar Bollywood ( )

T

- Train Simulator: Keisei, Toei Asakusa, Keikyu Line ( )
- Train Simulator: Kyushu Shinkansen ( )

W

- Wild Arms: Alter Code F ( )

X

- Xenosaga Episode I: Der Wille zur Macht ( )
- Xenosaga: Episode I Reloaded ( )

Y

- Yakuza 2 ( )

See also

- → List of PlayStation 2 games
- List of PlayStation 2 CD-ROM games
- List of PlayStation 2 network games
- List of Sony Greatest Hits games
- List of HD Enhanced PS2 games

List of PlayStation 2 games with HD support

The following is a **list of → PlayStation 2 games with support for HDTVs and EDTVs** as well as the games that have a 16:9 widescreen mode. Generally, progressive scan mode is activated by holding the Triangle and Cross, or "X," buttons down after the PlayStation 2 logo appears. When this is done, the game will typically load a screen with instructions on how to enable progressive scan. Many games only offer progressive scan through this method, offering no related options in the game's options menu. Both methods work on a PlayStation 3 as well. Component video cables are required for progressive scan mode.

When progressive mode is enabled on PAL (576i) games, the resolution is 480p, not 576p. Note that not all games from PAL territories support progressive scan mode 480p even if their NTSC U/C counterparts do.

A - F

→ Game	480p	16:9	1080i	480p on PAL
24: The Game	Yes	Yes	No	Yes
Ace Combat 04: Shattered Skies	No	Yes	No	No
Ace Combat 5: The Unsung War	No	Yes	No	No
Ace Combat Zero: The Belkan War	No	Yes	No	No
Alien Hominid (X+O)	Yes	No	No	Yes
Ape Escape 3	No	Yes	No	No
ATV Offroad Fury 2	Yes	No	No	No
ATV Offroad Fury 3	Yes	No	No	No
ATV Offroad Fury 4	Yes	No	No	No
Battlefield 2: Modern Combat	No	Yes	No	No
Beyond Good & Evil	Yes	Yes	No	Yes
Black	Yes	Yes	No	Yes
Blood Rayne 2	No	Yes	No	No
Bully	No	Yes	No	No
Burnout 3: Takedown	Yes	Yes	No	No
Burnout Dominator	Yes	Yes	No	No
Burnout: Revenge	Yes	Yes	No	Yes

Buzz! The Hollywood Quiz	No	Yes	No	No
Capcom Fighting Evolution	Yes	No	No	Yes
Cars	Yes	Yes	No	No
Cold Fear	No	Yes	No	No
Dance Dance Revolution X	Yes	No	No	No
Destroy All Humans	Yes	Yes	No	Yes
Destroy All Humans 2	Yes	Yes	No	No
Dragon Ball Z: Infinite World	Yes	No	No	No
Dragon Quest VIII	No	Yes	No	No
Driver: Parallel Lines	Yes	Yes	No	No
Family Feud	No	Yes	No	No
Final Fantasy XI	No	Yes	No	No
Final Fantasy XII	No	Yes	No	No

G - L

→ Game	480p	16:9	1080i	480p on PAL
Garou: Mark of the Wolves - NeoGeo Online [1]	Yes	Yes	No	No
Ghosthunter	Yes	Yes	No	Yes
Ghost in the Shell: Stand Alone Complex	Yes	Yes	No	No
Ghost Rider	No	Yes	No	No
God of War [6]	Yes	Yes	No	No
God of War II	Yes	Yes	No	No
Gran Turismo 3: A-Spec	No	Yes	No	No
Gran Turismo 4	Yes	Yes	Yes [2]	No
Grand Theft Auto III	No	Yes	No	No
Grand Theft Auto: Vice City	No	Yes	No	No
Grand Theft Auto: San Andreas	No	Yes	No	No
Grand Theft Auto: Liberty City Stories	No	Yes	No	No
Grand Theft Auto: Vice City Stories	No	Yes	No	No
Guilty Gear Isuka	Yes	No	No	No
Guilty Gear XX	Yes	No	No	No
Guilty Gear XX #Reload [3]	Yes	No	No	No
Guilty Gear XX Slash [1]	Yes	No	No	No
Guilty Gear XX Accent Core	Yes	No	No	No
Guilty Gear XX Accent Core PLUS	Yes	No	No	No
Guitar Hero II	Yes	Yes	No	Yes
Guitar Hero Encore: Rocks the 80s	Yes	Yes	No	Yes

Guitar Hero III: Legends of Rock	Yes	Yes	No	No
Guitar Hero Aerosmith	Yes	Yes	No	NO
Guitar Hero: Metallica	Yes	Yes	No	?
GUN	Yes	Yes	No	No
Gundam vs Zeta Gundam	Yes	No	No	No
Half-Life	No	Yes	No	No
Haunting Ground	Yes	No	No	Yes
Hitman 2: Silent Assassin	Yes	No	No	No
Hitman: Blood Money	Yes	No	No	Yes
I-Ninja	No	Yes	No	No
Incredible Hulk: Ultimate Destruction	Yes	Yes	No	No
Jak and Daxter: The Precursor Legacy	No	Yes	No	No
Jak II	Yes	Yes	No	Yes
Jak 3	Yes	Yes	No	Yes
Jak X: Combat Racing	Yes	Yes	No	Yes
Jet Li: Rise to Honor	Yes	No	No	No
Juiced	Yes	Yes	No	No
Kill Switch	Yes	Yes	No	No
Killzone	No	Yes	Yes	No
King of Fighters '98 Ultimate Match - NeoGeo Online	Yes	No	No	No
King of Fighters 2002	Yes	No	No	No
King of Fighters 2003	Yes	No	No	No
King of Fighters XI	Yes [1]	No	No	No
Legends of Wrestling II	Yes	No	No	No
Lego Star Wars 2: The Original Trilogy	No	Yes	No	No
Lemony Snicket's: A Series of Unfortunate Events	Yes	No	No	No
Lowrider	Yes	No	No	No

M - R

→ Game	480p	16:9	1080i	480p on PAL
Macross super dimensional fortress	No	Yes	No	No
Magna Carta: Tears of Blood	Yes	No	No	No
Mark of the Wolves	Yes	No	No	No
Mega Man X Command Mission	Yes	No	No	No
Mega Man X8	Yes	No	No	No
MLB 2004	Yes	No	No	No
MLB '06: The Show	Yes	No	No	No
MLB '07: The Show	Yes	Yes	No	No

MLB '09: The Show	No	Yes	No	No
Major League Baseball 2K5	Yes	No	No	No
Major League Baseball 2K6	Yes	No	No	No
Major League Baseball 2K7	Yes	Yes	No	No
Micro Machines V4	No	Yes	No	No
Monster Hunter	No	Yes	No	No
Mortal Kombat: Deception	Yes	Yes	No	No
Mortal Kombat: Armageddon	Yes	Yes	No	Yes
MVP Baseball 2004	Yes	No	No	No
MVP Baseball 2005	Yes	No	No	No
MX Unleashed	Yes	No	No	No
NBA 2K3	Yes	Yes	No	No
NBA 2K6	Yes	No	No	No
NBA 2K7	Yes	No	No	No
NBA 2K8	Yes	No	No	No
NBA 2K9	Yes	Yes	No	No
NBA Street Vol. 2	Yes	No	No	No
NCAA Football 2006	Yes	No	No	No
Need for Speed Carbon	No	Yes	No	No
Need for Speed Most Wanted	No	Yes	No	No
Need for Speed Underground 2	No	Yes	No	No
NeoGeo Battle Coliseum	Yes [1]	No	No	No
NFL 2K3	Yes	No	No	No
NFL 2K5	Yes	Yes	No	No
NHL 2K6	Yes	Yes	No	No
NiGHTS Into Dreams... [1]	Yes [3]	Yes	No	No
Outrun 2006: Coast to Coast	Yes	Yes	No	Yes
Outrun2SP: Special Tours	Yes	Yes	No	No
Pac-Man World Rally	Yes	Yes	No	No
Primal	Yes	Yes	No	Yes
Prince of Persia: The Sands of Time	Yes	No	No	Yes
Prince of Persia: The Two Thrones	Yes	No	No	Yes
Prince of Persia: Warrior Within	Yes	No	No	Yes
Project Zero 3	Yes	No	No	Yes
R: Racing Evolution	Yes [4]	Yes	No	No
Radiata Stories	Yes	Yes	No	No
Ratchet & Clank: Going Commando	Yes	Yes	No	No
Ratchet & Clank: Up Your Arsenal [6]	Yes	Yes	No	No

Ratchet: Deadlocked	Yes	Yes	No	No
Rayman 3: Hoodlum Havoc	Yes	Yes	No	Yes
Resident Evil 4	Yes	Yes	No	Yes
Robotech: Battlecry	Yes	No	No	No
Robotech Invasion	Yes	No	No	No
Rock Band [6]	Yes	Yes	No	??
Rock Band 2 [6]	Yes	Yes	No	??

S - Z

→ Game	480p	16:9	1080i	480p on PAL
Scarface: The World Is Yours	Yes	Yes	No	No
Sega Ages 20: Space Harrier II Complete Collection [1]	Yes	No	No	No
Sega Ages 24: Last Bronx ~Tokyo Bangaichi~ [1]	Yes	No	No	No
Sega Ages 25: Gunstar Heroes ~Treasure Box~ [1]	Yes	No	No	No
Sega Ages 28: Tetris Collection [1]	Yes	No	No	No
Sega Ages 29: Monster World Collection [1]	Yes	No	No	No
Sega Ages 30: Galaxy Force [1]	Yes	Yes	No	No
Sega Ages 32: Phantasy Star Complete Collection [1]	Yes	No	No	No
Sega Genesis Collection	Yes	Yes	No	No
Sega Rally 2006 [1]	No	Yes	No	No
Sega Rally Championship (bundled w/Sega Rally 2006) [1]	Yes	No	No	No
Sega Superstars Tennis	Yes	Yes	No	Yes
Shadow of the Colossus	Yes	Yes	No	Yes
Shadow the Hedgehog	Yes	No	No	No
Shadow of Rome	Yes	No	No	No
Shrek 2	Yes	No	No	No
Shox	No	Yes	No	No
Silent Hill Origins	No	Yes	No	No
Smash Court Tennis Pro Tournament 2	Yes	No	No	No
SOCOM: U.S. Navy SEALs	Yes	Yes	No	No
SOCOM II: U.S. Navy SEALs	Yes	Yes	No	No
SOCOM 3: U.S. Navy SEALs	Yes	Yes	No	No
SOCOM U.S. Navy SEALs: Combined Assault	Yes	Yes	No	No
Sonic Mega Collection Plus	Yes	No	No	No
Sonic Riders	Yes	No	No	No
Sonic Riders: Zero Gravity	Yes	No	No	Yes

Sonic Unleashed	Yes	No	No	No
Soul Calibur II	Yes	Yes	No	Yes
Soul Calibur III	Yes	Yes	No	No
Spawn: Armageddon	Yes	Yes	No	No
Splashdown: Rides Gone Wild	Yes	Yes	No	No
SSX 3	Yes	Yes	No	No
SSX on Tour	No	Yes	No	No
Star Ocean: Till the End of Time	Yes	Yes	No	No
Star Wars Bounty Hunter	Yes	No	No	No
Star Wars: The Force Unleashed	Yes	Yes	No	No
Street Fighter Alpha Anthology	Yes	No	No	Yes
Suikoden IV	Yes	No	No	Yes
Suikoden Tactics	Yes	No	No	No
SVC Chaos: SNK vs. Capcom	Yes	No	No	No
Syphon Filter: The Omega Strain	Yes	No	No	No
Syphon Filter: Dark Mirror	Yes	No	No	No
Tekken 4	Yes	No	No	Yes
Tekken 5	Yes	Yes	No	No
The Simpsons: Hit & Run	Yes	No	No	No
Thunder Force VI	Yes [5]	No	No	No
Tomb Raider Legend	Yes	Yes	No	Yes
Tomb Raider Anniversary	Yes	Yes	No	Yes
Tony Hawk's Pro Skater 4	No	Yes	No	No
Tony Hawk's Proving Ground	Yes	Yes	No	No
Tony Hawk's Project 8	Yes	Yes	No	No
Tony Hawk's American Wasteland	Yes	Yes	No	No
Tourist Trophy	Yes	Yes	Yes [2]	No
True Crime: Streets of LA	Yes	Yes	No	Yes
True Crime: New York City	Yes	Yes	No	No
UFC: Throwdown	Yes	No	No	No
Urban Reign	Yes	Yes	No	No
Vampire Darkstalkers Collection [1]	Yes	No	No	No
Vampire Night	Yes	No	No	No
Valkyrie Profile 2: Silmeria	Yes	Yes	No	No
Van Helsing	Yes	Yes	No	No
Vexx	Yes	No	No	No
Way of the Samurai	Yes	No	No	No
Yakuza	No	Yes	No	No
Yakuza 2	No	Yes	No	No

Notes

- Note 1: Only in NTSC-J version.
- Note 2: Only in NTSC-J and NTSC-U/C versions.
- Note 3: Not released in NTSC-U/C.
- Note 4: Only in NTSC-U/C.
- Note 5: Not compatible with PlayStation 3.
- Note 6: No Reference. Can change these settings in option menu.

References

- "High Definition Game Database [1]". http://hdgames.net. Retrieved 2007-01-01.
- HDTV Arcade PlayStation 2 database [2]

See also

- → List of PlayStation 2 games
- List of Xbox games with HD support

References

[1] http://hdgames.net

[2] http://www.hdtvarcade.com/hdtvforum/index.php?autocom=custom&page=ps2ab

Article Sources and Contributors

PlayStation 2 *Source*: http://en.wikipedia.org/w/index.php?title=PlayStation_2 *Contributors*: $tinkOman, -Majestic-, 07lancer, 2D, 2mcm, 343 Guilty Conscience, 4RM0, 4finger, 5369Mary, 94vian, A Man In Black, A Nobody, A strolling player, ABF, AR Argon, ARC Gritt, ASHTONZANECKI, Abce2, Abeg92, Abhisara, Ableadded, Abrech, Ace of Sevens, Action Jackson IV, Adaobi, Adashiel, AdjustShift, Adot93, AdviceKid123, Aero Leviathan, After Midnight, Ahoerstemeier, AimalCool, Aitias, Aksi great, Alansohn, Alchemy101, Aldie, AlexOvShaolin, Alexf, Alexsutton, Alexwcovington, AlistairMcMillan, AliveFreeHappy, Allen649, Alphax, Alucard 16, Aman27deep, Ameltzer, Anclation, Andersekbom, Andi Saleh, AndonicO, AndrewCrogonklol, Andrewpmk, Andy, Andypandy.UK, Anetode, AngelHedgie, AngelOfSadness, Angela, Angelic Wraith, Anger22, Angr, AniMate, Anonymous 82, Antandrus, Anthall1991, Antonio Lopez, Apetc, Apostrophe, Apparition11, Aravindkannankara, Arctic-Editor, Aresmo, ArielGold, Aristotle ManBearPig, Arjun01, Arslan4life, Arthur2045, Arwel Parry, Asc99c, Asdquefty, Asher196, Assntit3sr4fuNN, AstroPig7, Auric, Aussie Evil, AxG, AxelBoldt, Axeman89, Ayrton Prost, B, B1link82, BD2412, Bachrach44, Backpackadam, BalazsH, Ballaholic Baller 123456789, Bardeep7, Barek, Bassman91, Batbump, Bawolff, Bdhoff, Beardy10, Beeglebug, Beginning, Bencey, Benwildeboer, Beta, Bezking, Bfleisher, Bharat411, Black Reign 56, Blades, Blake-, Blanchardb, BloodDoll, Blucheese, Blue Laser, Blue520, BlueDevil, BlueMint, Bluekka, Blugu128, Bmicomp, Boarder8925, Bobblewik, Bobby 67897474758, Bobo192, BobtheVila, Bobthebobthetimothy, Bongwarrior, Bonus Onus, BorgQueen, Boshaman92, Bowserv2, BrOnXbOmBr21, BradBeattie, Bradley28, Brahulprasad, Brazil4Linux, Brett 91091, Brucelee, Bryan Derksen, BryanG, Bsadowski1, Bubberbob, Bucketsofg, Buffalobob41, BurnShot, ByteofKnowledge, C trillos, C'est moi, C0nanPayne, CBM, CTUFieldOpsDirector, CWY2190, Cagado, CaitSithX, Calamity jones, CalusReyma, Camw, Can't sleep, clown will eat me, CanisRufus, Capecodeph, Capricorn42, Captain Disdain, Captain panda, Carbonox-infernox, CardinalDan, Carey Evans, Carl Adamson, Carnildo, Casper Gutman, Casper10, CastAStone, Caster23, Casull, CatMan, Catgut, Cburnett, Celarnor, Cenarium, Centrx, Chadlupkes, ChaosMaster, Chapultepec, Chase me ladies, I'm the Cavalry, Chaser, Chavando, Chealer, Cheawybean55, Cheesecakemonstar, Chensiyuan, ChibiKuririn, Chicken7, Chivafighter93, Chmod007, Chowbok, Chtito, Ciao 90, Cinder6, Ck lostsword, Clngre, Closedmouth, Cloudwolf124, Cobra 3000, Coffee Atoms, Col. Hauler, Colbane117, Colonthree, Combination, CommonsDelinker, Comradeash, Con 017, Consolegames, Consumed Crustacean, Conversion script, Cookiecaper, Cool3, Coolquin, Corbin 654, Cpc464, Cpeterson12, Crackerbelly, Crazycomputers, Crazyfrengy, Creidieki, Cremepuff222, Crockster, Crumbsucker, CrunchyCookie, CryptoDerk, Csaag, Csus814, Ctjf83, Culverin, Curps, Curran1980, CyberMew, CyberSkull, Cyfal, D0762, D607, DCEvoCE, DChiuch, Dabomb87, Daev, Damian Yerrick, Dan Rode, Dan100, Dancter, Daniel Case, Danny69(. Y .), Danski14, DarkFalls, Darkhunger, Darranc, Darth Panda, Darukaru, Dat11, DaveJB, Davelong, David011, Dawn Bard, Dctcool, DeABREU, DeadEyeArrow, Deathawk, Deathtrap3000, Debug-GED, Deckiller, Dedication of Hearts, Delahe15, Delta 91, DennyColt, Deor, DerHexer, Deusfaux, Devindred, Dg0896, DiabloJones, Diagonalfish, Dibbity Dan, Diceman, Diefor, Dionyseus, Diretiva, Disastrophe, Discopriest, Discospinster, Dispenser, Dlae, Dlohcierekim's sock, DmitryKo, Dngo18, Docbml, Doctorinthahouse, Doniago, Doom jester, Doom127, DoubleCross, Douglasr007, Dp462090, Dr.alf, DragonWR12LB, Dragonballmaster26, Draicone, Drain222000, Drat, DrewT2, Drilnoth, Dro0001, Drtangles, Dtaylor1984, Dtremenak, Dubkiller, Dues Ex Machina, Dungeon Siege, Duomillia, ESkog, EVIL-C44, Eadsmasha, Ed g2s, EddEdmondson, Edgar181, Editor99433, Editore99, Edward321, Effer, ElTyrant, ElderScrollsFan, Eldude611, Electricmoose, Elfguy, EliasAlucard, Elkman, Elliskev, Eng101, Ennerk, Enviroboy, Epbr123, Eried, Escape Orbit, EscapingLife, Estel, Estoy Aquí, Ethylene Kid, Eurocave, Everyking, Evice, Evil Monkey, Excirial, Ezgamer, F, Falcon866, Falcon9x5, Faradayplank, Faragon, Feinoha, Ffgamera, Fieldday-sunday, Finalnight, Firedragon2133, Firsfron, Flabby72, Flax9, FlyingPenguins, Fnagaton, Fnfd, Foudre, Fourohfour, Fracture91, Frasierfreak, Frasor, Freakofnurture, Frecklefoot, Frederlck, Fredrik, FreplySpang, Freyr, Frightner, Froth, FrumpyTheClown, FullMetal Falcon, Furrykef, Fyyer, G4M2 Betty, GT4GTR, Gadfium, Gadse, Gail, Gaius Cornelius, Galoubet, Gamepolice, Gamzsone V2, Ganryuu, Garden, GatesPlusPlus, Gatoatigrado, Gatorademan, Gatta, Gbeeker, Gdo01, GeminiDomino, Gerbrant, Getcrunk, Gh87, Ghostkid11, Gimmetrow, Gleffler, Glen, Goatasaur, God of War, Godzson123, Gogeta324, Gogo Dodo, Golbez, GoldDragon, Gormanly, Grandgrawper, GreatWhiteNortherner, GregAsche, Grenavitar, Greyarea0972, Grika, GrimRepr39, GroundZero, Grubber, Gtamonster, Gtdp, Gurch, Gurchzilla, Gutwrenchtat, Guybrush, Guyinblack25, Gwendibbley, Gürkan Sengün, HUNTERSYU, Haakon, Hadal, Hagrid's half brother, Hairy Dude, HalfShadow, Halo 01010101, Halsteadk, Hanakosama, Hapenis, Happinessiseasy, Happy piggy, Harej, Hariharan91, Harryboyles, Haseo9999, Hashar, Hateless, Havok, Hayabusa future, Hayden120, Haza-w, Head, HeavyD14, Hede2000, Hellbus, Hellcat fighter, Hello32020, Hemanshu, Hertz1888, HexaChord, Hibana, Hijak, Hillcoco12, Hinotori, Hirokazu, Hitflip39, Hofac07, Hohonn, Horncomposer, Hotnikelz, Hrld171, Ht1848, Hubscratch, Husond, Hut 8.5, Hydrogen Iodide, IAMTHEEGGMAN, IMNOTAVANDALK5, IRT.BMT.IND, Iain99, Ianjones1900, Ianjones50, Ibagli, Ibanez RYM, Icarus3, Iced Kola, Idleguy, Iggy402, Igordebraga, Ilia ilia ilia, Illiniball64, Ilovepenis, Imapianoman, ImperatorExercitus, InvaderJim42, InvisibleK, Ipino, Iridescent, Ishmaelblues, Isiahjacks, Ixfd64, J Di, J S Firefox, J. Nguyen, J.delanoy, JForget, JONJONAUG, JSpung, JTBX, Jacek Kendysz, Jacj, Jack fa el do, Jackel, Jagamijs, Jagzthebest, Jaileer, JakRox, James P Twomey, James henry Lee, James086, Japan\'s flag, Japanese biplane, Jason One, Jastein, Java13690, Javier Donoso, Jaxl, Jbond311, Jboyle4eva, Jdlyall, Jdmdreamz94, Jecates, Jedi6, Jeff3000, Jeffrey Mall, Jeffrey Smith, Jeremy Visser, Jeromey55555, Jhouser24, JiMidnite, Jigahurtz, Jiggelmaster7, Jim Anonym, Jim Douglas, Jim.Liu, Jinnai, Jj137, Jkbena612, Jklin, Jmcollier, Joaquin546, JoeSmack, John Mash, John Reaves, John254, Jon Harald Søby, Jonnyboy911, JorgeGG, Jorvik, Joyous!, Jrcplanet, Jrdioko, Jtalledo, Juppiter, Justinfr, Jzcool, K1Bond007, K1darkknight, K1sso, KBBMessi, Kafziel, Kapn Korea, Kappa, KathrynLybarger, Kazikame, Kbolino, Keilana, Keithustus, Kelvin.gck1978, Kelvinc, Kencaesi, Kevin Breitenstein, Khukri, Kicking222, Kid56, King of Hearts, Kingpin13, Kingsin, Kinless, KitAlexHarrison, KittenKiller, KnowledgeOfSelf, Knowledgeman66, Konstable, Koweja, Kowloonese, Krakalaken, Krellis, Krich, Krustybrandmaf, Krystyn Dominik, KsprayDad, Kungfuadam, Kuru, Kusma, Kuzaar, Kylee20051, Kylu, LAX, La Pianista, Ladsgroup, Lafraia, LancerEvolutionMR, Lanpaper, LaptopGun, Large incisors, Larsinio, Latinoboy3, Latka, Lavenderbunny, Lawshmaw, Lcarscad, LeaveSleaves, Lectonar, LedgendGamer, Lee Cremeans, Leedeth, Leewilson123, Leflyman, Legend78, LeoNomis, Leodj1992, Lethalwiki, Lexi Marie, Liftarn, Lightmouse, Linuxlad, Linuxrules1337, Little Mountain 5, Longhair, Loopman12, Lord Hawk, Loren.wilton, Lowellian, Lradrama, Luboogers25, Lucozade 93, LudBob, Luk, Luna Santin, Lunchscale, Luph25, M Johnson, M1ss1ontomars2k4, MASTERuser, MBisanz, MER-C, MK8, MZMcBride, MaNeMeBasat, Macaddct1984, Macara, Madcow 93, Majorly, Malcohol, Malcolm, Maleous, Malinaccier, Malo, Manop, Marc Mongenet, Marcan, Marek69, MarioV, Markintellect, Martin451, Martinp23, Marvelvsdc, Master Thief Garrett, MasterJag, Matboy, Matias.Reccius, Matteh, Matthew.daniels, Mattygabe, MaxMahem, Maxwahrhaftig, Mboverload, McGeddon, McSly, Mcbridelr, Mcjakeqcool, Me676, Mech80, Mechasheherezada, Meeples, Megakhfanbej2, Megbucko, Mehrunes Dagon, Memanmo, Mence Master, Mentifisto, MeowMixer, Merope, Merovingian, Message From Xenu, Mewtu, Mfield, Michael Farris, Michael.Pohoreski, Michaeltriola, Michal Nebyla, MickWest, Microsoft Fanboy, Mifter, Mike Dillon, Mike Rosoft, Mike Yaloski, Mike1, Mikeblas, Mikon8er, Milkfish, Minesweeper.007, Miquonranger03, Misterjta, Mixwell, Mjg3456789, Mmendezs, Modulatum, Mohammedaqeel9, Momus, Momusufan, Montydrei, MoogleEXE, Mooglemoogle, Mooncowboy, Moonriddengirl, Mooshmooshman, MordredKLB, Morio, Morton devonshire, Movementarian, Mpvide65, Mr toasty, Mr. Lefty, Mr. Pointy, Mr.Do!, Mr.NorCal55, Mr.Z-man, Mr2001, MrBubbles, Mrbojangles657, Mrwojo, Mrzaius, Muchosucko, Munkyinvasion, Mushroom, Musse-kloge, Mvent2, Myanw, Myscrnnm, NE2, NHS2008, NYyankees51, Nakon, Narson, Natalie Erin, NateDan, NawlinWiki, Needlenose, Neil Kemp, Neji255, NekoFever, Nemesis88@wikipedia.org, Neonumbers, Neuro, New Age Retro Hippie, Nick125, NickBush24, Nicolas Capens, Nihiltres, Nintendude, Nishkid64, Nixeagle, Norm, Notheruser, Nrbelex, NuclearWarfare, Nuggetboy, Nv8200p, Nyenyec, Nyjil, Nyletak, OST63BIOS, Ochib, Offensiveandconfusing, Old Guard, Oliver Lineham, OllieFury, Omicronpersei8, Only1kanobi, Oooo, Optichan, OranL, Orderinchaos, Orjanlothe, Ormers, Orzomn, Oscarthecat, Osgar12, Otduff, Otherlleft, Otisjimmy1, Ottawa4ever, Outburn, Outlyer, Oxymoron83, P08.gratton, P924s88, PDH, PJ Pete, PKFC, PS2lover, PS3 wins, PSTC555, PSXer, PaPiRiCoSuAvE, Painofsense04, Pak21, Panny 666, PaperTruths, Para, Parallel or Together?, Pardthemonster, Paris By Night, Passive, Patjennings45, Paul August, Paul and carol xd, Paul levesque, Payam81, Pb30, Pd THOR, Pedantic of Purley, Pedracer400, Pennywisepeter, Perfect Proposal, Perks, Peruvianllama, Peter Winnberg, Peter5floor, Pg2114, Pgk, Phediuk, PhilHibbs, Philip Trueman, Philthecow, Phosphorous, PhotoBox, Photoguy439, Piano non troppo, Piecemealcranky, Pielover87, Pikawil, Pilot 51, Pilotguy, Pip2andahalf, Pipotchi, Piroroadkill, Plornt, Plugwash, Pogogunner, Poiuyt Man, Polaralex, Polluks, Poomachine, Pooporama, Poor Yorick, Poorboy, Porqin, Potatofan52, PrestonH, Prodego, Professional Gamer, Pseudomonas, Puchiko, Putnamehere3145, Qk, Quackshot, Quadell, Quiddity, Qutezuce, Qwe, R'n'B, RC-0722, RKN1987, RPH, Rada, Radagast, RadicalBender, Ragnarok Addict, Ragster128, RainbowOfLight, Rairaunak, Raist3d, Rajrajmarley, Ramirososa07, Randomfartguy, Randy Rotman, Raptordrew, Rattlesnake, Raveb, Raven4x4x, RazorICE, ReallyMale, RedHillian, Reddi, Redquark, Redvers, Reedy, Remaen, RememberMe?, Renegadeviking, Requires1GB, Res2216firestar, Reticulated, Rettetast, RevolverOcelotX, RexNL, ReyBrujo, Rhobite, Rholton, Ribeirowaldo, Rich1000, RichJohnCena, Richiekim, Rick Browser, RickK, Ricky540, Rilak, Ringbang, Rjd0060, Rjwilmsi, Rl, Rlevse, RoWhite, RobJ1981, Robert John Scott, Robertotr, Robertvan1, Robivy64, Robmods, Robomaeyhem, RobyWayne, Rockysmile11, Roger Davies, Rohawn, Rolken, Roodog2k, Rory096, Rory77, RoyBoy, Rror, Rsciaccio, Rtyq2, Runtime, Ryanrossiter, Ryndgmn, S.Skinner, S3000, S33k3r, SEJohnston, SG-17, SJP, SNEST2, SNIyer12, SNS, Sabenben, Saddhiyama, Sade, Salavat, Salmar, Sam Hocevar, Sam Korn, Sam Li, SamBishop, SamuraiFez, Sandahl, Sanhern, Sapp Krupter, Satoru003, Saturncast, Saxbryn, Scapler, Scepia, Sceptre, Scetoaux, SchnitzelMannGreek, SchumiChamp, Scientizzle, Sciurinæ, Sdfisher, Sean William, Seba5618, Seidenstud, Seph VII, Sertan, Serversurfer, Sesshomaru, Sfweg8oy8fwa734oghasn93pvn2789p3r4y189, Shadow Hog, Shadow assist, Shadow187187, ShadowInferno, ShadowLaguna, Shadowjams, Shadownet 46, Shadypalm88, Shamanstk, Shark64, Shawnc, Shawnhath, Shifter95, Shizane, ShoopWoop, Shorty5220, Shotwell, Sietse Snel, Sihabuddin, Silly rabbit, Sillygostly, Silver Edge, Silver Sonic Shadow, SilverWerewolf, Simply south, SiobhanHansa, Sixteen Left, Sko5000, Skor, Skv avenger, Sky Attacker, SkyBon, SkyWalker, Slowking Man, Smack, Smarterthanu91, Smashedupps2, Snowolf, SoWhy, Soccerpro, Sockatume, Soetermans, Song4758, Sonu27, Soosed, Sosekopp, Sparrowman980, Spazure, Special-T, Species8473, Spinna94, Splash, Sportsfan101, Squater, Srolf3, Ssolbergj, Staffwaterboy, Starblind, Stay, Steel, Stej01, Stephenb, Steve jha, Steven 2005, Stick it to the man, Stiii, Stormie, Stuart P. Bentley, Studtrooper, Sully76cl, Supernick345, Superway25, Sushiflinger, Swordsman04, Sylent, Syrthiss, TJ Spyke, TKD, Takeo12361, TakuyaMurata, Tartarus, Tawker, Taxman, Tayoyaki, Technolust, Teggles, Tehw1k1, Teresa44, Terrapin, Texture, Tfl, Tgraham87, Th1rt3en, Thaddius, The Anome, The Arc, The Hemp Necktie, The Rambling Man, The luigi kart assasions, The-bus, TheDotGamer, TheKoG, TheListUpdater, TheN1Armyguy, TheSuave, TheXenocide, TheYmode, Theaveng, Theface102, Thejumperkin, Theman00, Themat21III, Thetanmancan, Thingg, Third Strike, Thomas PL, Thomas.macmillan, ThorSkaagi, Thorpe, Thumperward, Thunderbird2, Thunderbrand, Thurls, Tide rolls, TigerShark, Tigerred53, Tj9991, TjOeNeR, Tjansen, Tlim7882, Tmandrake13, Tmckellar, Tnyqu, Toastyman, Tobyc75, Tocharianne, Tohd8BohaithuGh1, Tomkinsc, Tommo92, TonicBH, Tonius, Tonkatsu182, Tony22r, Tonytam, Tregoweth, Trevorthemachomanevans, TreyGeek, Tripacer99, Truflip99, Tschel, Tubedogg, TutterMouse, Twister Twist, TylerMad8, Typhoon, URORIN, Ubik70, Ugur Basak, Ukexpat, Ultraflame, Uncle Dick, Uniracer45, Unit 1010, Unknownwarrior33, Until It Sleeps, Useight, User27091, Userguy70, V4run Agarwal, Vague Rant, Valeriya, Valoem, VanHalen, Vandalizor07, Vdlsnes, Vdub49, Veemonkamiya, Veledan, Versus22, Vertius, Vinibinini, Vinnyv, Viper82, ViperSnake151, Vishnava, Viskonsas, Visor, Vlad, Vulcanstar6, Vvidetta, W Tanoto, WODUP, WadeSimMiser, Wafulz, Warfieldian, Warmpuppy2, Warthawg, Wasted Sapience, WatermelonPotion, Whale plane, Where, WhisperToMe, Whsn, WikHead, Wiki alf, Wiki b00b12, Wiki cleanupteam, Wikio475, WikipedianMarlith, Wikiuser15, Wilbern Cobb, Willking1979, Wilt, Wimt, WinterSpw, Wizzard, Wootery, Wordbuilder, WulfTheSaxon, Wwagner, Wweisreal, Wysprgr2005, X!, X1987x, X201, XFD, XMODXX, XMORPHEUSX, XXShigaXX, Xaosflux, Xeno, Xezbeth, Xizer, Xp54321, Xu3w3nan, Xy7, Y control, Y2kcrazyjoker4, Yamaguchi先生, Yamamoto Ichiro, Yanksox, Yeanold Viskersenn, Yengkit19, Yipdw, Ynot544, Yoda2287, YolanCh, Yuckfoo, Yuri Elite, Z.E.R.O., Zabadoh, Zacharychung, ZakuSage, Zapdsl, ZeldaX, Zephyr2k, Zer0efx, Zero1328, ZeroRaider, Zeta26, Zidane4028, Zigger, Zilog Jones, ZimZalaBim, Zreeon, Zzuuzz, 2702 anonymous edits

PSX (DVR) *Source*: http://en.wikipedia.org/w/index.php?title=PSX_%28DVR%29 *Contributors*: 2help, After Midnight, Akadewboy, Angela, Anthony Appleyard, Arminius, Ashitaka96, AutoFire, BAxelrod, Bartledan, Belinrahs, Blakegripling ph, Bluesxxxman, Bmicomp, Brazil4Linux, BrianKnez, CBM, Can't sleep, clown will eat me, CanisRufus, Centrx, Chairman S., Ciao 90, Closetoeuphoria, Cmdrjameson, Crackeh, CyberSach, CyberSkull, DabMachine, Dancter, DanielPenfield, Darth Mike, Deor, Djr xi, Doom127, Draky, Dsd510, Dtcdthingy, ESkog, Ed g2s, Effer, Enviroboy, Estel, Ex-Nintendo Employee, FaithLehaneTheVampireSlayer, GTA Ganxtaize, Gabriellundmark, Ganryuu, Ganymead, Garden, Gene Nygaard, Gspawn, Gurchzilla, HalfShadow, Ht1848, Imroy, Indon, J.delanoy, J4lambert, JadziaLover, Jem112, Jeremy Visser, Jigahurtz, John254, Jtalledo, Justindube, Jzhang, K1Bond007, Kicking222, Krash, Krystyn Dominik, Kungming2, Lazi, LeCire, Lightmouse, LonelyWolf, Lord Hawk, Lushsight, Main Event, Matt Britt, Melloss, Mika1h, Mild Bill Hiccup, Millard73, Mintleaf, Mitaphane, Mushroom, Myscrnnm, N00body, N5iln, NP Chilla, NeonMerlin, NickBush24, Oliverdl, Oni Lukos, Oroso, PKFC, Paracropolis, ParticleMan, Pastorsboy, Pfkninenines, PhilHibbs, Piroroadkill, Plutoniumboss, Quadell, RJASE1, RJaguar3, Ragnarok Addict, Randext, Randomfartguy, Rawerxxxaw, RayAYang, Red, Renaissancee, Rettetast, ReyBrujo, Rick Browser, RoyBoy, ST47, Samafito, Scepia, SchumiChamp, Secfan, Shawnhath, ShelfSkewed, Shinigami Josh, Silver Edge, Skier Dude, Sonicspike41, Ssolbergj, Stratadrake, Sum233, SupaGorilla306, SuperDude115, TechGuyDude9X, Th1rt3en, That Guy, From That Show!, TheEnlightened, Thingg, Thue, Tobz1000, Trega123, TubularWorld, Until It Sleeps, W Tanoto, WadeSimMiser, Wangi, Warmpuppy2, Wiki alf, Work permit, Xtreme racer, ZetaReveal, 241 anonymous edits

DualShock *Source*: http://en.wikipedia.org/w/index.php?title=DualShock *Contributors*: -Majestic-, 21655, AVRS, Abrech, Ace Class Shadow, Aeon1006, Ajuk, Aknorals, Albrozdude, Alex43223, Alexander5959, AmiDaniel, Amren, Anetode, Any door, Armaced, Astatine-210, Atirage, Atlant, Barnolde, BeefJeaunt, Bhavesh.Chauhan, Bladestorm, Bleako, Bloigen, Boarder8925, Bobdoe, Bollinger, Bootatahpotato, Brazil4Linux, Briandiaz, Bwatts28, CASE, Cadsuane Melaidhrin, CalumH93, Calvins48, Can't sleep, clown will eat me, CaseyPenk, Cheeseman Muncher, Cholmes75, Chowbok, Chuckiesdad, Ciao 90, ClarinetFreak28, Cliché Online, Consumed Crustacean, Cormier6083, Crazycomputers, Cris Spiegel, Crumb, CyberSkull, DBZROCKS, DaSauce, Damian Yerrick, Dancter, Darkdoom3000, Darkhunger, Darranc, Darthkenobi, Deadman1848, Deathawk, Discospinster, Dispenser, Doom127, Dragonscales, ENeville, Ed g2s, Einstein the afrodude, EliasAlucard, Enviroboy, Epbr123, Erencexor, Escobar4Life, Ex-Nintendo Employee, Falcon9x5, Frap, Fratrep, Furrykef, G026r, Gamer007, Geg, GodzillaX8, GrandDrake, Greensburger, Grouf, Hailey C. Shannon, Hairy Dude, Halofan.3, Hamushka11, Hateless, Hbdragon88, Hello2112, Hrimfaxi, Hyad, Ibanez RYM, Imroy, Iridescent, JaffaCakeLover, Jason One, JeloMulawin, Jheinz, Jj sexy, Jlhflex, Jordan Brown, Jzhang, K1Bond007, KabutoHunter, Kappa, Katalaveno, Keyser Söze, Kieff, Kinglink, Kord, Kozuch, Lanky, Letdorf, Little Professor, Luismbs, MIT Trekkie, MTX, Machete97, Malig123, MegaLinkv2, Metal Sonic v2.0, Mitaphane, Mitsukai, Mockenoff, Momus, Moonriddengirl, Mushroom, Myscrnnm, NatureA16, Neodarksaver, NickBush24, Nintendude, Ogmo6492, Oliveg1, Onetwo1, Oniamien, Oscarthecat, Paper Luigi, Patriotic dissent, Phinn, PipChaos, Pissant, Poirier273, Ponyo, Project10509, PureLegend, Pyrosim, Ragnarok Addict, Raiker, Rajb245, Rajrajmarley, Ravensfan5252, Redlioness, ReyBrujo, RockMFR, Royboy5371, Sam Hocevar, Schi, Schizogony, ShadowedBlade, Shawnc, ShelfSkewed, Silver Edge, Sixteen Left, Smack, Smoothy, Sockatume, SomeStranger, Spamguy, SparksBoy, Spoonboy42, Ssolbergj, Steel angel kurumi, Stratadrake, Supertone, SynergyBlades, TJ Spyke, Teac77, Th1rt3en, Thaddius, The Haunted Angel, The93owner, TheFreshPrinceOfBel-Air, Thingg, Tirkfl, Titan Lad, Tobycat, Tobyo1, Tofof, Tom, Travelbird, Tsumng, Tubedogg, TubularWorld, Tyan23, Vague Rant, Valepert, Victorgrigas, Werthog, Whsn, Wisq, Wonchop, X201, Xizer, Yellowdesk, Zoganes, 390 anonymous edits

PlayStation 2 Expansion Bay *Source*: http://en.wikipedia.org/w/index.php?title=PlayStation_2_Expansion_Bay *Contributors*: A Nobody, Brazil4Linux, Brian Kendig, Cross stix, DMG413, Ennerk, Hibana, Ht1848, Hymyly, IRT.BMT.IND, Invitatious, JVz, JaGa, KJS77, Kappa, Katefan0, Linuxlad, Lord Hawk, Mickyfitz13, Nargalzius, Niteowlneils, Nuttycoconut, One, PKFC, Palm9x, PiaCarrot, Psa-, Quasipalm, Richi902, Rick Browser, SamuraiClinton, Shadow demon, Someguy1221, TerokNor, Th1rt3en, Tubedogg, Vague Rant, Wapcaplet, Y2kcrazyjoker4, Zidane2k1, Zilog Jones, 56 anonymous edits

EyeToy *Source*: http://en.wikipedia.org/w/index.php?title=EyeToy *Contributors*: A Man In Black, A Nobody, A. B., Ableadded, Adambro, AeronPeryton, Aldenis, Alynna Kasmira, AndyDent, Antandrus, Axem Titanium, Bachrach44, Back ache, Benwildeboer, Blehfu, Blueliteway, Brent01, Brian Kendig, Brianna Goldberg, Brownspank, Burton Radons, Canderson7, Chris Sharman, Chrislk02, Cookiecaper, D.brodale, D2htornado, DaProx, Dancter, Davidkazuhiro, DeadEyeArrow, Deathawk, Deltabeignet, Dispenser, Dysprosia, Elven6, Enviroboy, Evercat, Feldhamer, Feureau, Fl, Fleminra, Fr4zer, Fubaz, Gcm, Geljamin, Generica, Georgepowell2008, Glocks Out, Gtg204y, GunterPete, Haradona, Harris-Grad, HeavyD14, Hibana, Hooperbloob, Ht1848, Ilikemusic, Ixtli, J-Ros, J.delanoy, JP Godfrey, JSellers0, Jagripino, Jakob.kalt, Jason One, JiFish, Jinnantonnix, Kappa, Kariteh, Kowloonese, Kurt Jansson, Lord Hawk, Madskunk, Markthemac, Mausy5043, Mentifisto, Mo0, Modster, Monkey14, Mr toasty, Mrsapple, MudkipNDS, Myanw, Myscrnnm, N. Harmonik, N0rbie, NSH002, Nancy, Neurolysis, New Age Retro Hippie, Norm, Optichan, Paris By Night, Pavel Vozenilek, Persian Poet Gal, Preada, PsychoJosh, Puchiko, Quaque, RadioActive, Ragefury32, RattleMan, Rb, RedRollerskate, Reinoutr, Russell Freeman, Sbisolo, Scepia, Scottie, Segafreak2, Silver Edge, Starwind Amada, Stuartmanderson, SynergyBlades, TJ Spyke, Th1rt3en, The Rogue Penguin, The Storm Surfer, Thebdogg, Themasterofwiki, Thorpe, TicketMan, Tigerghost, TimR, Tntnnbltn, TreyGeek, Triforce of Power, Trisreed, Unyoyega, Vanieter, Vcelloho, VikC, ViperSnake151, Whumit, WikHead, Wknight94, X201, Xevious, Xtreme racer, Y2kcrazyjoker4, Zoganes, 273 anonymous edits

PlayStation 2 Headset *Source*: http://en.wikipedia.org/w/index.php?title=PlayStation_2_Headset *Contributors*: ADeveria, After Midnight, Baseballfan, Bovineboy2008, Dragon's Light, Fiftyquid, Fishhead, GMR-ROX, Hibana, Iain Cheyne, K1Bond007, Mr toasty, Myscrnnm, Open2universe, Randall Brackett, Reaper X, Remuel, Rick Browser, Seidenstud, Sock of Steel359, Th1rt3en, Tubedogg, Zoganes, 37 anonymous edits

Logitech *Source*: http://en.wikipedia.org/w/index.php?title=Logitech *Contributors*: -Majestic-, 1dragon, A Nobody, Achilles2.0, Achromatic, Adambonneruk, Adric, Aeons, Aitias, Anas Salloum, Ancjr, Avala, Avjoska, Avllr, Benlisquare, Benutzer123, Bergsten, Berkut, Bhavesh.Chauhan, Binksternet, Bleh999, Bob200607, Bobby1928, Bongwarrior, Bovineone, BrianRecchia, Can't sleep, clown will eat me, Canderson7, Carbonferum, Catgut, Changster308, Chealer, CheatLemur, Chont, Chris G, Closeapple, Coolcaesar, Corporal clegg48, Crazytales, Cs92, Cspurrier, CutterX, DChiuch, DMG413, Daram.G, Darin-0, Darrensavery, Darth NormaN, Docu, Dondon0, Dpbsmith, Dvdkdrmn, Dysepsion, Dysprosia, Edmarriner, Emx, Enlil Ninlil, Erc, Escapeartist, Ethanmarcus, Excirial, Exeunt, Finlay McWalter, Flakeloaf, Fosnez, Frencheigh, Fsdfs, Funkerbunks, Geekfox, Gertlex, Gestumblindi, Glane23, Gmarini, Gogo Dodo, GregorB, Grutness, Gtennis, Gurch, HTurtle, Hadriven, Haseo9999, Hc5duke, HollyAm, Hurricane111, JForget, Jacob Poon, Jamcib, Jan eissfeldt, Jawed, Jclemens, Jdabney, Jdrice8, Jeanjpoirier, Jefe2000, Jmcc150, Jmlk17, John Fader, Jovianeye, Kelly Martin, Kemiv, Kevinchze, Kimon, KjtheDj, Kocio, Krogoth Zero, Kuru, Kyle1278, Leonbev, Lexein, Lights, Lonaowna, Lousyd, Lypheklub, MMuzammils, MacintoshWriter, Malepheasant, Manix, Marc Mongenet, Mardus, Mastercaster, Mattym129, Memodude, Mendaliv, Metaphoria, Mhiley, Mikeblas, Mikecron, Mistsrider, Mitch Ames, Mitchomaniac, Mlewis000, Modeha, Modster, Moeron, MrOllie, Myscrnnm, N328KF, Neenish Tart, Njan, Norm, OHFM, OSborn, Pekayer11, PhilFree, Philip Trueman, Platypus222, Pollenberg, Posix memalign, Pularoid, R0pe-196, RadicalBender, Rake, Rama, Redvers, Retro00064, ReyBrujo, Richard W.M. Jones, RobScheurwater, Robguru, Rodan44, Ronhjones, Salgueiro, Samel tvom, Samsara, SebastianHelm, Segaba, ShadowHntr, Sherool, SkyWalker, Slo-mo, Splintax, Sprinter76, Spyco, Starcraft101, SteinbDJ, Stephane.magnenat, Synergy, THE KING, TedE, The undertow, TheBilly, TheTim1997, Thelawns, These7enthprophet, Thibbs, TinFoil, Tkgd2007, Tokek, Tyciol, Ulric1313, Vague Rant, Voice of All, Voyagerfan5761, Waryklingon, Wengero, Wii25, Wik, Wiki-vr, Witchwooder, Wiz126, Woohp978, WyldStallionRyder, XGM, Xmnemonic, ZeroOne, 195 anonymous edits

Linux for PlayStation 2 *Source*: http://en.wikipedia.org/w/index.php?title=Linux_for_PlayStation_2 *Contributors*: 10nitro, 16@r, Academic Challenger, Acroterion, Admrboltz, Antaeus Feldspar, Binarybits, Bollinger, Brazil4Linux, Bucephalus, Chealer, Ciao 90, Colin Keigher, D'Agosta, Debug-GED, Ed g2s, ElfMage, Erencexor, Frap, Ganryuu, Gurch, Guy2007, InShaneee, Interrobang², Invitatious, Kesla, Kross, Laurens, Lavenderbunny, Lemmer1800, Lord Hawk, Macanima, Masnevets, Mentifisto, Michael963, MichaelAhlers, Mike Richardson, Minghong, Myscrnnm, Ndboy, Oscarthecat, PyroGamer, Radiant chains, RadioActive, Ragnarok Addict, Roadstaa, Rob*, Sherlockindo, Sietse Snel, Tedernst, Thrane, Thumperward, Tothwolf, TreyGeek, Vegaswikian, 57 anonymous edits

PCSX2 *Source*: http://en.wikipedia.org/w/index.php?title=PCSX2 *Contributors*: Aaru Bui, AbbasJin, Aillema, Allexx54, AnaTo, Arite, Bayo, Blakegripling ph, Bryan121, CesarB, ConZZor, CyberSkull, Dancter, Dark Shikari, Dashboardy, Dungodung, EconomicsGuy, Epheuu, Erencexor, Evilgohan2, Figjam88au, FironDraak, General Plot, Haseo9999, Heinrich717, Ilion2, Isuldor, J. Nguyen, K1Bond007, Kbdank71, Lightmouse, Little Professor, Lone Guardian, Lusht, Matariel, Mdkcheatz, Meiskam, Michael Drüing, Mohamedhp, Mossman93, Mr toasty, Mrf, N. Harmonik, NateDan, Oni Lukos, Onsmelly, OverlordQ, Playstationman, Polluks, ReCover, Renegadeviking, Rvalles, Sandebert, Screenshotguy, Sd31415, ShaunMacPherson, SilentChasm, Sn0wflake, Snarius, Stewartadcock, TheParanoidOne, Thunderbrand, Toehead2001, TomTheHand, Usucdik2, Vindictive Warrior, Ween the diff, Winhunter, Zwmalone, 123 anonymous edits

List of PlayStation 2 games *Source*: http://en.wikipedia.org/w/index.php?title=List_of_PlayStation_2_games *Contributors*: 141, 2fort5r, 604555infofreak, A More Perfect Onion, ADS190, AMK152, Abdull, Ablenet, Acetic Acid, AdventurerGR, Aero X360, AeronPeryton, Agadomichael, Ahoerstemeier, AlexOvShaolin, Allan64, Alynna Kasmira, Amigobro2, Analoguedragon, Anetode, Angie Y., AngoraFish, Angusmclellan, Antriksh Yadav, Apnicholson7, Apostrophe, Arienh4, Art LaPella, Aspects, Axem Titanium, Bdashti98, Belasted, BigRedMF, Biolizard, Bjarner, Blades, BlahBlahBlah124, Blazingluke, Blink182dude720, Blueliteway, Bobblehead, Bobo192, Bovineboy2008, Brandizzo, Brazil4Linux, Brett 91091, Brian0918, Bulbabean, Bykergrove, C777, CF90, CO, CWii, Can't sleep, clown will eat me, Captainthrust, Carlos De Los Santos, CasanovaUnlimited, Casper10, Celithemis, Cenarium, Ceros, Ceyockey, Champishere21, Charles Matthews, Chicago119, Chocobo ff, Chored, Chris Ssk, Chris1219, Clamticore, Coasttocoast, Colt9033, Combination, ConantheLibrarian, Corn Popper, Crimsonfox, CryptoDerk, Cs-wolves, Cubs Fan, Cun, CyberSkull, Cyberdemon007, Cyrus XIII, D6, DINOMAN, DOAsaturn, DT29, Damien Russell, Dang08, DangApricot, DannyB!!, Darklock, Dawg1279, Dbreakey, Deadkid dk, Deejjj, Defsac, Dell9300, Delmothop, Deltabeignet, Dinohunter999, Distraction, DocDragon, Doctor33, DocumentN, Dragonfly888, DraxusD, Drister117gunit, Drkirby, Dude867, Dylgr, ELF, Ed g2s, Edge o Matic 2005, Edokter, Eliashc, EmanWilm, Empty2005, Ennerk, Eric42, Etyek, Evan1975, EveryDayJoe45, FCFordlord, FPWR, FatKidsJiggle, Fbv65edel, Federelli, Feeder18, Feinoha, Felix Wan, Fenikkusuzero, Filpaul, Fratrep, Frecklefoot, Fromnothertime, FutureNJGov, Gadfium, Gaius Cornelius, GameLegend, Gamer5545, Gangster26, Garavello, Gatogirl, Gh87, Gilliam, Ginsublender, Gohmifune, Gorgack2000, Gracefool, Great Legacy, GrimRepr39, Gtrmp, Guinea pig warrior, Gurch, Guyinblack25, Haipa Doragon, Hall Monitor, Hamdrew, Hardworker111, Hbent, Helopticor, Hibana, Historiographer, Homestar14, Hothguard11, Ht1848, Hughey, Hymeer, Iain k, Ianblair23, IcemanAD, Ikram000, ImanAtwal, Imre Toth, Ixfd64, JIP, Jae millz, JamesHoadley, Jappalang, Jason One, Jasonglchu, Jef-Infojef, Jeff G., Jeff3000, JeffW, Jerec, JimRhodes, Jimmiejohnson2008, JI931, Jmayer, JohannVisagie, JohnTheGamer, JohnnyMrNinja, Jonathan D. Parshall, Juhachi, K1Bond007, KConWiki, Kablammo, Kafuffle, KahranRamsus, Kappa, Kawnhr, Kctsang, Khym Chanur, KillingForCulture, King Wagga, KingZog, Kingerik, Kirbysuperstar, Kogoro 9 23, Kona1611, KramarDanIkabu, Kronnang Dunn, Kylemcauliffe15, LGagnon, LadyofShalott, Lasttan, Lcarscad, LevelNth, Lightlowemon, Lightmouse, Lightsup55, Linczone, Lochdog11, Lockshaw13, LordOfAllGaming, LtHija, Luked, Luna Santin, M2ger, M8gen, MIT Trekkie, MTC, Magioladitis, Majin Izlude, Malcolmxl5, Marasmusine, MarcK, Mariofan90, Markles, Martinman11, Masem, Matt0401, Matthewrules, Mattyatty, Mattyo007, Mausy5043, Maverick Leonhart, Mcconnellwade, Mcjakeqcool, Mclarenaustralia, Megaman en m, Mgp2005es, Michael Devore, Mighty Mongoose, Mika1h, Mike902, Mintguy, Misza13, Mmxx, Modica40, Mohammedaqeel9, Momus,

Monkeyblue, MoonShadow, Mr pand, Mr. Garrison, Mr.Radzilla, Mrsteak613, Mrtreats, Muchi, Mwalimu59, Mygerardromance, Myrdred, N. Harmonik, NSH002, NeoChaosX, Netoholic, New Age Retro Hippie, New User, Nihonjoe, Nintendax, Nintendofan5000, Norm, Oakster, Open2universe, OriginalJunglist, Orjanlothe, Ortolan88, Oscarthecat, Otrfan, PV250X, Patternott, Paul A, Pegship, Penale52, Peripitus, Phantz, Phil1988, PhilMT311, Pieboyjr, Pikawil, Playa26, Plugwash, Poiuyt Man, Presea6934, Professional Gamer, Project23, Prophet15, Pruneau, RaccoonFox, RadicalBender, Radicool, Radiokid1010, Raiden29o9, Ramkumari, Rcjsuen, RedWordSmith, RememberMe, Remurmur, Rev Tie Dye, Richiekim, RickoniX, Rickus Muller, Riffsan, Rjwilmsi, RobJ1981, Robina Fox, Robinepowell, Rocketgoat, Rockin56, Rodparkes, Rory77, Rossami, Ryan Roos, Ryan the Game Master, Ryanasaurus0077, Salavat, Sam Burke, Sar1192, Sasuke Kid, Sb2007, Seigi Choujin, Seryass, ShadowKinght, ShelfSkewed, Shrek32, Sidney Gould, Silver Edge, Sim0n20, Simon12, SimonP, Simonkoldyk, Sinnyo, Slo-mo, Smomo, Sock of Steel359, Someone another, SoundPound500000, Spider-Man, Ssjgoku75x, St Fan, Staticz, Steel, Stef Nighthawk, Steve, Stormie, Strike Chaos, Supertrinko, Swerdnaneb, SwordKirby537, TBK0000, TEHCHUCKLEBROTHERS, TJ Spyke, TKD, Tabletop, TangentCube, Tassedethe, Tbone762, Teacher of madness, Template namespace initialisation script, Teniii, Th1rt3en, The Irons, The Tramp, The-, TheDotGamer, Thebrid, Third Strike, Thorpe, Thricecube, Thunderbrand, Tictac1098, Tim!, Timkovski, Tkanasen, TomEatsCake, TomSon, TrafficBenBoy, TrapStilton, TraxPlayer, Trivialist, Tromatic, Tubedogg, Twistedkombat, Tyman 101, U6 pebl, Unisouth, Unknown566, User27091, V3rt1g0, Vanguard, Vanky, Vear, Veilor, Versus22, Vetes, Vikifanatic, Vitoraraujo, Vitz-RS, WOSlinker, Wadester5, Walesdbr, Wandering Ghost, Welsh, WhereAmI, Woohookitty, Workster, Wtshymanski, X201, Xevious, Xezbeth, Xino, Yandman, Yer0c, ZJP, ZS, ZeWrestler, Zenithian, ZeroOne, Zeta26, Zoganes, Zoggie50, 1264 anonymous edits

List of PlayStation 2 DVD-9 games *Source*: http://en.wikipedia.org/w/index.php?title=List_of_PlayStation_2_DVD-9_games *Contributors*: Adaobi, Bkessler23, Computerjoe, Dp462090, EddieVanZant, EditingMachine, Eeeeh, Federelli, Kcgreatfox, Kitten cannon lover, Marasmusine, N. Harmonik, Ragnarok Addict, Rory77, Shadwolf, Silver Edge, Sjakkalle, Snesmaster40, Tonyrayo, Trivialist, Zidane4028, 97 anonymous edits

List of PlayStation 2 games with HD support *Source*: http://en.wikipedia.org/w/index.php?title=List_of_PlayStation_2_games_with_HD_support *Contributors*: Akadewboy, Akatsuki.sousui, Al2k4, Alex43223, AlphaTwo, Archen, BRPXQZME, Bmecoli, Btnheazy03, Daram.G, Devindred, Drunkenmonkey, Eusis, Geniac, Gh87, Goyingus, Gui874111, Guppy313, Iaoth, Ixnay, Izarate, JAF1970, Jackman197, Jason One, Jkid4, Jol123, JtMinahan, Judzillah, Kona1611, Konsolkongen, LtHija, Marasmusine, Marshall tindall, Mfs95, Millionairex, Minuzzo 69, Mittio, Mr.bonus, NeoChaosX, Nnickn, Onesimos, Paris By Night, Pearle, Pumapayam, Redawgts, Rolod, Sd31415, Shawnc, Silhouette of Mushroom, Silver Edge, SkiDragon, Slipeth, Snakelda, Somno, Takeheed, The-G-Unit-Boss, Thebrid, Thekoolaidpimp, Tziouv, WikHead, Woch, Yellow Mage, Ykzb, Yworo, Zenlax, Zippedpinhead, 334 anonymous edits

Image Sources, Licenses and Contributors

File:PlayStation 2 logo.svg *Source*: http://en.wikipedia.org/w/index.php?title=File:PlayStation_2_logo.svg *License*: unknown *Contributors*: Rjd0060, Ssolbergj

File:PlayStation 2.png *Source*: http://en.wikipedia.org/w/index.php?title=File:PlayStation_2.png *License*: GNU Free Documentation License *Contributors*: Original uploader was Java13690 at en.wikipedia Later versions were uploaded by Asim18 at en.wikipedia.

Image:Memory Card for PlayStation 2.jpg *Source*: http://en.wikipedia.org/w/index.php?title=File:Memory_Card_for_PlayStation_2.jpg *License*: GNU Free Documentation License *Contributors*: User:Qurren

Image:Sony Dual Shock 2.jpg *Source*: http://en.wikipedia.org/w/index.php?title=File:Sony_Dual_Shock_2.jpg *License*: unknown *Contributors*: Original uploader was Boarder8925 at en.wikipedia

File:PlayStation 2 Remote Control.jpg *Source*: http://en.wikipedia.org/w/index.php?title=File:PlayStation_2_Remote_Control.jpg *License*: unknown *Contributors*: User:Gürkan Sengün

File:Playstation 2.jpg *Source*: http://en.wikipedia.org/w/index.php?title=File:Playstation_2.jpg *License*: Public Domain *Contributors*: Asim18, Dancter, NeMeSiS

File:SCPH-75000CB.jpg *Source*: http://en.wikipedia.org/w/index.php?title=File:SCPH-75000CB.jpg *License*: GNU Free Documentation License *Contributors*: User:Qurren

File:PS2 PS@ slimline front IGP3831.jpg *Source*: http://en.wikipedia.org/w/index.php?title=File:PS2_PS@_slimline_front_IGP3831.jpg *License*: GNU Free Documentation License *Contributors*: Kozuch, Photoguy439

File:PS2Slim.JPG *Source*: http://en.wikipedia.org/w/index.php?title=File:PS2Slim.JPG *License*: GNU Free Documentation License *Contributors*: Bayo, Dancter, Frumpy, 2 anonymous edits

File:Ps215467890 IGP3832.jpg *Source*: http://en.wikipedia.org/w/index.php?title=File:Ps215467890_IGP3832.jpg *License*: GNU Free Documentation License *Contributors*: Kozuch, Photoguy439, Sushiflinger

File:PS2 IGP3830.JPG *Source*: http://en.wikipedia.org/w/index.php?title=File:PS2_IGP3830.JPG *License*: GNU Free Documentation License *Contributors*: Kozuch, Photoguy439, Sushiflinger

File:Slim silver ps2.JPG *Source*: http://en.wikipedia.org/w/index.php?title=File:Slim_silver_ps2.JPG *License*: Public Domain *Contributors*: Robivy64 (talk). Original uploader was Robivy64 at en.wikipedia

File:EyeToy.JPG *Source*: http://en.wikipedia.org/w/index.php?title=File:EyeToy.JPG *License*: GNU Free Documentation License *Contributors*: Bayo, Dancter, Str4nd, 1 anonymous edits

File:Resident evil 4 chainsaw controller.jpg *Source*: http://en.wikipedia.org/w/index.php?title=File:Resident_evil_4_chainsaw_controller.jpg *License*: Public Domain *Contributors*: Original uploader was Le Grand Roi des Citrouilles at en.wikipedia

File:Sony EmotionEngine CXD9615GB top.jpg *Source*: http://en.wikipedia.org/w/index.php?title=File:Sony_EmotionEngine_CXD9615GB_top.jpg *License*: GNU Free Documentation License *Contributors*: Dancter, EugeneZelenko, GreyCat, Morkork, Qurren

File:Sony Graphics Synthesizer CXD29314GB.jpg *Source*: http://en.wikipedia.org/w/index.php?title=File:Sony_Graphics_Synthesizer_CXD29314GB.jpg *License*: Public Domain *Contributors*: Benctwu, Dancter, Idrougge

File:SCPH39000-GS.JPG *Source*: http://en.wikipedia.org/w/index.php?title=File:SCPH39000-GS.JPG *License*: Public Domain *Contributors*: User:FEAR6655

File:CXD9833GB.png *Source*: http://en.wikipedia.org/w/index.php?title=File:CXD9833GB.png *License*: unknown *Contributors*: User:Offensiveandconfusing

File:Neweegs.JPG *Source*: http://en.wikipedia.org/w/index.php?title=File:Neweegs.JPG *License*: Public Domain *Contributors*: User:Robivy64

File:Sony Playstation 1 CPU.jpg *Source*: http://en.wikipedia.org/w/index.php?title=File:Sony_Playstation_1_CPU.jpg *License*: Public Domain *Contributors*: Benctwu, Dancter

Image:Psxlogo.svg *Source*: http://en.wikipedia.org/w/index.php?title=File:Psxlogo.svg *License*: unknown *Contributors*: User:Ciao 90

Image:Console psx.jpg *Source*: http://en.wikipedia.org/w/index.php?title=File:Console_psx.jpg *License*: Public Domain *Contributors*: Moi

Image:Dvd xmb img02 pop.jpg *Source*: http://en.wikipedia.org/w/index.php?title=File:Dvd_xmb_img02_pop.jpg *License*: unknown *Contributors*: After Midnight, Fetofs, Imroy, Jusjih, N00body, Rhopkins8, Rick Browser, VMS Mosaic, 4 anonymous edits

Image:Sonydualshockblack.JPG *Source*: http://en.wikipedia.org/w/index.php?title=File:Sonydualshockblack.JPG *License*: unknown *Contributors*: User:Onetwo1

Image:PS2 ExpansionBay.jpg *Source*: http://en.wikipedia.org/w/index.php?title=File:PS2_ExpansionBay.jpg *License*: unknown *Contributors*: User:PiaCarrot

Image:NTSC-UC HDD Package.JPG *Source*: http://en.wikipedia.org/w/index.php?title=File:NTSC-UC_HDD_Package.JPG *License*: unknown *Contributors*: User:Richi902

Image:PS2 NetworkAdapter.jpg *Source*: http://en.wikipedia.org/w/index.php?title=File:PS2_NetworkAdapter.jpg *License*: unknown *Contributors*: User:PiaCarrot

Image:PlayStation BB Unit.jpg *Source*: http://en.wikipedia.org/w/index.php?title=File:PlayStation_BB_Unit.jpg *License*: unknown *Contributors*: User:PiaCarrot

Image:Eyetoy logo.svg *Source*: http://en.wikipedia.org/w/index.php?title=File:Eyetoy_logo.svg *License*: unknown *Contributors*: Neurolysis, Tntnnbltn

Image:EyeToy.JPG *Source*: http://en.wikipedia.org/w/index.php?title=File:EyeToy.JPG *License*: GNU Free Documentation License *Contributors*: Bayo, Dancter, Str4nd, 1 anonymous edits

Image:Logitech logo.png *Source*: http://en.wikipedia.org/w/index.php?title=File:Logitech_logo.png *License*: logo *Contributors*: Platypus222

Image:Logitechheadquarters.jpg *Source*: http://en.wikipedia.org/w/index.php?title=File:Logitechheadquarters.jpg *License*: GNU Free Documentation License *Contributors*: Coolcaesar, Ejfetters

Image:Logitech-gamepad.JPG *Source*: http://en.wikipedia.org/w/index.php?title=File:Logitech-gamepad.JPG *License*: Public Domain *Contributors*: Bayo, Mardus, Samel tvom

Image:Ps2 linux kit setup.jpg *Source*: http://en.wikipedia.org/w/index.php?title=File:Ps2_linux_kit_setup.jpg *License*: Creative Commons Attribution 2.5 *Contributors*: Original uploader was Colin Keigher at en.wikipedia

Image:Ps2 linux contents.jpg *Source*: http://en.wikipedia.org/w/index.php?title=File:Ps2_linux_contents.jpg *License*: Creative Commons Attribution 2.5 *Contributors*: Original uploader was Colin Keigher at en.wikipedia

Image:Ps2 linux dvd.jpg *Source*: http://en.wikipedia.org/w/index.php?title=File:Ps2_linux_dvd.jpg *License*: Creative Commons Attribution 2.5 *Contributors*: Colin Keigher, Liftarn, N. Harmonik, Project FMF

Image:pcsx2 logo.png *Source*: http://en.wikipedia.org/w/index.php?title=File:Pcsx2_logo.png *License*: unknown *Contributors*: Aaru Bui, Toehead2001

Image:PCSX2.png *Source*: http://en.wikipedia.org/w/index.php?title=File:PCSX2.png *License*: GNU General Public License *Contributors*: PCSX2 Team

Image:Flag of Europe.svg *Source*: http://en.wikipedia.org/w/index.php?title=File:Flag_of_Europe.svg *License*: Public Domain *Contributors*: User:-xfi-, User:Dbenbenn, User:Funakoshi, User:Jeltz, User:Nightstallion, User:Paddu, User:Verdy p, User:Zscout370

Image:Flag of the United States.svg *Source*: http://en.wikipedia.org/w/index.php?title=File:Flag_of_the_United_States.svg *License*: Public Domain *Contributors*: User:Dbenbenn, User:Indolences, User:Jacobolus, User:Technion, User:Zscout370

Image:Flag of Japan.svg *Source*: http://en.wikipedia.org/w/index.php?title=File:Flag_of_Japan.svg *License*: Public Domain *Contributors*: Various

GNU Free Documentation License Version 1.2, November 2002 Copyright (C) 2000,2001,2002 Free Software Foundation, Inc. 59 Temple Place, Suite 330, Boston, MA 02111-1307 USA Everyone is permitted to copy and distribute verbatim copies of this license document, but changing it is not allowed.

0. PREAMBLE

The purpose of this License is to make a manual, textbook, or other functional and useful document "free" in the sense of freedom: to assure everyone the effective freedom to copy and redistribute it, with or without modifying it, either commercially or noncommercially. Secondarily, this License preserves for the author and publisher a way to get credit for their work, while not being considered responsible for modifications made by others. This License is a kind of "copyleft", which means that derivative works of the document must themselves be free in the same sense. It complements the GNU General Public License, which is a copyleft license designed for free software. We have designed this License in order to use it for manuals for free software, because free software needs free documentation: a free program should come with manuals providing the same freedoms that the software does. But this License is not limited to software manuals; it can be used for any textual work, regardless of subject matter or whether it is published as a printed book. We recommend this License principally for works whose purpose is instruction or reference.

1. APPLICABILITY AND DEFINITIONS

This License applies to any manual or other work, in any medium, that contains a notice placed by the copyright holder saying it can be distributed under the terms of this License. Such a notice grants a world-wide, royalty-free license, unlimited in duration, to use that work under the conditions stated herein. The "Document", below, refers to any such manual or work. Any member of the public is a licensee, and is addressed as "you". You accept the license if you copy, modify or distribute the work in a way requiring permission under copyright law. A "Modified Version" of the Document means any work containing the Document or a portion of it, either copied verbatim, or with modifications and/or translated into another language. A "Secondary Section" is a named appendix or a front-matter section of the Document that deals exclusively with the relationship of the publishers or authors of the Document to the Document's overall subject (or to related matters) and contains nothing that could fall directly within that overall subject. (Thus, if the Document is in part a textbook of mathematics, a Secondary Section may not explain any mathematics.) The relationship could be a matter of historical connection with the subject or with related matters, or of legal, commercial, philosophical, ethical or political position regarding them. The "Invariant Sections" are certain Secondary Sections whose titles are designated, as being those of Invariant Sections, in the notice that says that the Document is released under this License. If a section does not fit the above definition of Secondary then it is not allowed to be designated as Invariant. The Document may contain zero Invariant Sections. If the Document does not identify any Invariant Sections then there are none. The "Cover Texts" are certain short passages of text that are listed, as Front-Cover Texts or Back-Cover Texts, in the notice that says that the Document is released under this License. A Front-Cover Text may be at most 5 words, and a Back-Cover Text may be at most 25 words. A "Transparent" copy of the Document means a machine-readable copy, represented in a format whose specification is available to the general public, that is suitable for revising the document straightforwardly with generic text editors or (for images composed of pixels) generic paint programs or (for drawings) some widely available drawing editor, and that is suitable for input to text formatters or for automatic translation to a variety of formats suitable for input to text formatters. A copy made in an otherwise Transparent file format whose markup, or absence of markup, has been arranged to thwart or discourage subsequent modification by readers is not Transparent. An image format is not Transparent if used for any substantial amount of text. A copy that is not "Transparent" is called "Opaque". Examples of suitable formats for Transparent copies include plain ASCII without markup, Texinfo input format, LaTeX input format, SGML or XML using a publicly available DTD, and standard-conforming simple HTML, PostScript or PDF designed for human modification. Examples of transparent image formats include PNG, XCF and JPG. Opaque formats include proprietary formats that can be read and edited only by proprietary word processors, SGML or XML for which the DTD and/or processing tools are not generally available, and the machine-generated HTML, PostScript or PDF produced by some word processors for output purposes only. The "Title Page" means, for a printed book, the title page itself, plus such following pages as are needed to hold, legibly, the material this License requires to appear in the title page. For works in formats which do not have any title page as such, "Title Page" means the text near the most prominent appearance of the work's title, preceding the beginning of the body of the text. A section "Entitled XYZ" means a named subunit of the Document whose title either is precisely XYZ or contains XYZ in parentheses following text that translates XYZ in another language. (Here XYZ stands for a specific section name mentioned below, such as "Acknowledgements", "Dedications", "Endorsements", or "History".) To "Preserve the Title" of such a section when you modify the Document means that it remains a section "Entitled XYZ" according to this definition. The Document may include Warranty Disclaimers next to the notice which states that this License applies to the Document. These Warranty Disclaimers are considered to be included by reference in this License, but only as regards disclaiming warranties: any other implication that these Warranty Disclaimers may have is void and has no effect on the meaning of this License.

2. VERBATIM COPYING

You may copy and distribute the Document in any medium, either commercially or noncommercially, provided that this License, the copyright notices, and the license notice saying this License applies to the Document are reproduced in all copies, and that you add no other conditions whatsoever to those of this License. You may not use technical measures to obstruct or control the reading or further copying of the copies you make or distribute. However, you may accept compensation in exchange for copies. If you distribute a large enough number of copies you must also follow the conditions in section 3. You may also lend copies, under the same conditions stated above, and you may publicly display copies.

3. COPYING IN QUANTITY

If you publish printed copies (or copies in media that commonly have printed covers) of the Document, numbering more than 100, and the Document's license notice requires Cover Texts, you must enclose the copies in covers that carry, clearly and legibly, all these Cover Texts: Front-Cover Texts on the front cover, and Back-Cover Texts on the back cover. Both covers must also clearly and legibly identify you as the publisher of these copies. The front cover must present the full title with all words of the title equally prominent and visible. You may add other material on the covers in addition. Copying with changes limited to the covers, as long as they preserve the title of the Document and satisfy these conditions, can be treated as verbatim copying in other respects. If the required texts for either cover are too voluminous to fit legibly, you should put the first ones listed (as many as fit reasonably) on the actual cover, and continue the rest onto adjacent pages. If you publish or distribute Opaque copies of the Document numbering more than 100, you must either include a machine-readable Transparent copy along with each Opaque copy, or state in or with each Opaque copy a computer-network location from which the general network-using public has access to download using public-standard network protocols a complete Transparent copy of the Document, free of added material. If you use the latter option, you must take reasonably prudent steps, when you begin distribution of Opaque copies in quantity, to ensure that this Transparent copy will remain thus accessible at the stated location until at least one year after the last time you distribute an Opaque copy (directly or through your agents or retailers) of that edition to the public. It is requested, but not required, that you contact the authors of the Document well before redistributing any large number of copies, to give them a chance to provide you with an updated version of the Document.

4. MODIFICATIONS

You may copy and distribute a Modified Version of the Document under the conditions of sections 2 and 3 above, provided that you release the Modified Version under precisely this License, with the Modified Version filling the role of the Document, thus licensing distribution and modification of the Modified Version to whoever possesses a copy of it. In addition, you must do these things in the Modified Version: A. Use in the Title Page (and on the covers, if any) a title distinct from that of the Document, and from those of previous versions (which should, if there were any, be listed in the History section of the Document). You may use the same title as a previous version if the original publisher of that version gives permission. B. List on the Title Page, as authors, one or more persons or entities responsible for authorship of the modifications in the Modified Version, together with at least five of the principal authors of the Document (all of its principal authors, if it has fewer than five), unless they release you from this requirement. C. State on the Title page the name of the publisher of the Modified Version, as the publisher. D. Preserve all the copyright notices of the Document. E. Add an appropriate copyright notice for your modifications adjacent to the other copyright notices. F. Include, immediately after the copyright notices, a license notice giving the public permission to use the Modified Version under the terms of this License, in the form shown in the Addendum below. G. Preserve in that license notice the full lists of Invariant Sections and required Cover Texts given in the Document's license notice. H. Include an unaltered copy of this License. I. Preserve the section Entitled "History", Preserve its Title, and add to it an item stating at least the title, year, new authors, and publisher of the Modified Version as given on the Title Page. If there is no section Entitled "History" in the Document, create one stating the title, year, authors, and publisher of the Document as given on its Title Page, then add an item describing the Modified Version as stated in the previous sentence. J. Preserve the network location, if any, given in the Document for public access to a Transparent copy of the Document, and likewise the network locations given in the Document for previous versions it was based on. These may be placed in the "History" section. You may omit a network location for a work that was published at least four years before the Document itself, or if the original publisher of the version it refers to gives permission. K. For any section Entitled "Acknowledgements" or "Dedications", Preserve the Title of the section, and preserve in the section all the substance and tone of each of the contributor acknowledgements and/or dedications given therein. L. Preserve all the Invariant Sections of the Document, unaltered in their text and in their titles. Section numbers or the equivalent are not considered part of the section titles. M. Delete any section Entitled "Endorsements". Such a section may not be included in the Modified Version. N. Do not retitle any existing section to be Entitled "Endorsements" or to conflict in title with any Invariant Section. O. Preserve any Warranty Disclaimers. If the Modified Version includes new front-matter sections or appendices that qualify as Secondary Sections and contain no material copied from the Document, you may at your option designate some or all of these sections as invariant. To do this, add their titles to the list of Invariant Sections in the Modified Version's license notice. These titles must be distinct from any other section titles. You may add a section Entitled "Endorsements", provided it contains nothing but endorsements of your Modified Version by various parties--for example, statements of peer review or that the text has been approved by an organization as the authoritative definition of a standard. You may add a passage of up to five words as a Front-Cover Text, and a passage of up to 25 words as a Back-Cover Text, to the end of the list of Cover Texts in the Modified Version. Only one passage of Front-Cover Text and one of Back-Cover Text may be added by (or through arrangements made by) any one entity. If the Document already includes a cover text for the same cover, previously added by you or by arrangement made by the same entity you are acting on behalf of, you may not add another; but you may replace the old one, on explicit permission from th previous publisher that added the old one. The author(s) ar publisher(s) of the Document do not by this License giv permission to use their names for publicity for or to assert or imp endorsement of any Modified Version.

5. COMBINING DOCUMENTS

You may combine the Document with other documents release under this License, under the terms defined in section 4 above f modified versions, provided that you include in the combination of the Invariant Sections of all of the original document unmodified, and list them all as Invariant Sections of yo combined work in its license notice, and that you preserve all the Warranty Disclaimers. The combined work need only contain or copy of this License, and multiple identical Invariant Sections ma be replaced with a single copy. If there are multiple Invaria Sections with the same name but different contents, make th title of each such section unique by adding at the end of it, parentheses, the name of the original author or publisher of tha section if known, or else a unique number. Make the sam adjustment to the section titles in the list of Invariant Sections the license notice of the combined work. In the combination, yo must combine any sections Entitled "History" in the variou original documents, forming one section Entitled "History likewise combine any sections Entitled "Acknowledgements", an any sections Entitled "Dedications". You must delete all section Entitled "Endorsements".

6. COLLECTIONS OF DOCUMENTS

You may make a collection consisting of the Document and oth documents released under this License, and replace th individual copies of this License in the various documents with single copy that is included in the collection, provided that yo follow the rules of this License for verbatim copying of each of th documents in all other respects. You may extract a sing document from such a collection, and distribute it individual under this License, provided you insert a copy of this License in the extracted document, and follow this License in all oth respects regarding verbatim copying of that document.

7. AGGREGATION WITH INDEPENDENT WORKS

A compilation of the Document or its derivatives with oth separate and independent documents or works, in or on a volum of a storage or distribution medium, is called an "aggregate" if th copyright resulting from the compilation is not used to limit th legal rights of the compilation's users beyond what the individua works permit. When the Document is included in an aggregat this License does not apply to the other works in the aggregat which are not themselves derivative works of the Document. If th Cover Text requirement of section 3 is applicable to these copie of the Document, then if the Document is less than one half of th entire aggregate, the Document's Cover Texts may be placed o covers that bracket the Document within the aggregate, or th electronic equivalent of covers if the Document is in electron form. Otherwise they must appear on printed covers that bracke the whole aggregate.

8. TRANSLATION

Translation is considered a kind of modification, so you ma distribute translations of the Document under the terms of sectio 4. Replacing Invariant Sections with translations requires specia permission from their copyright holders, but you may includ translations of some or all Invariant Sections in addition to th original versions of these Invariant Sections. You may include translation of this License, and all the license notices in th Document, and any Warranty Disclaimers, provided that you als include the original English version of this License and th original versions of those notices and disclaimers. In case of disagreement between the translation and the original version this License or a notice or disclaimer, the original version w prevail. If a section in the Document is Entitle "Acknowledgements", "Dedications", or "History", the requiremer (section 4) to Preserve its Title (section 1) will typically requir changing the actual title.

9. TERMINATION

You may not copy, modify, sublicense, or distribute th Document except as expressly provided for under this Licens Any other attempt to copy, modify, sublicense or distribute th Document is void, and will automatically terminate your righ under this License. However, parties who have received copies or rights, from you under this License will not have their license terminated so long as such parties remain in full compliance.

10. FUTURE REVISIONS OF THIS LICENSE

The Free Software Foundation may publish new, revise versions of the GNU Free Documentation License from time t time. Such new versions will be similar in spirit to the preser version, but may differ in detail to address new problems concerns. See http://www.gnu.org/copyleft/. Each version of th License is given a distinguishing version number. If the Documer specifies that a particular numbered version of this License "c any later version" applies to it, you have the option of followin the terms and conditions either of that specified version or of an later version that has been published (not as a draft) by the Fre Software Foundation. If the Document does not specify a versio number of this License, you may choose any version eve published (not as a draft) by the Free Software Foundatior ADDENDUM: How to use this License for your documents To us this License in a document you have written, include a copy of th License in the document and put the following copyright an license notices just after the title page: Copyright (c) YEAR YOU NAME. Permission is granted to copy, distribute and/or modif this document under the terms of the GNU Free Documentatio License, Version 1.2 or any later version published by the Fre Software Foundation; with no Invariant Sections, no Front-Cove Texts, and no Back-Cover Texts. A copy of the license is include in the section entitled "GNU Free Documentation License". If yo have Invariant Sections, Front-Cover Texts and Back-Cove Texts, replace the "with...Texts." line with this: with the Invarian Sections being LIST THEIR TITLES, with the Front-Cover Text being LIST, and with the Back-Cover Texts being LIST. If yo have Invariant Sections without Cover Texts, or some othe combination of the three, merge those two alternatives to suit th situation. If your document contains nontrivial examples c program code, we recommend releasing these examples i parallel under your choice of free software license, such as the GNU General Public License, to permit their use in free software.

CPSIA information can be obtained at www.ICGtesting.com
Printed in the USA
LVOW071713200212

269551LV00012B/139/P